DISCOVERING

DOORSTEP WILDLIFE

DISCOVERING

DOORSTEP WILDLIFE

A Complete Guide to the Animals and Plants of Towns, Parks and Gardens

JOHN FELTWELL

HAMLYN

Photographic Acknowledgements

KEY: Centre = c; Left = l; Right = r; Top = t; Bottom = b
All the photographs used in this book were supplied by John Feltwell/Wildlife Matters, except:

Heather Angel/Biofotos *(pages 25b, 38-39, 130r)*; Ardea/R J C Blewitt *(page 43r)*; Biofotos *(page 110bl)*; John Buckingham/Wildlife Matters *(page 42l)*; Derek Budd/Wildlife Matters *(page 116r)*; Bruce Coleman/ S C Potter *(page 46br)*; Robin Fletcher *(pages 60l, 101 br)*; Jacana/ Ducrot *(page 82tl)*; Jacana/Dupont *(page 31r)*; Jacana/G Sommer *(page 99r)*; G Kinns/Biofotos *(page 93t)*; Frank Lane/N Schzempp *(page 16r)*; Pat Morris/Wildlife Matters *(pages 11l, 34b, 41r, 49l, 50l, 51l, 53b, 66r, 76t)*; NHPA/Stephen Dalton *(page 101b)*; NHPA/ G E Hyde *(pages 89t, 110bc)*; NHPA/K G Preston-Mafham *(page 110br)*; NHPA/Werner Zeps *(page 86r)*; Don Taylor/Wildlife Matters *(page 16t)*; Wildlife Studios Limited *(page 19t)*; M W F Tweedie/Wildlife Matters *(page 35l, 35r, 37t)*.

Illustration Acknowledgements

Sarah De Ath – 34-35, 61, 71, 80, 95, 117; John Gosler *– 12 (left), 20-21, 68-69, 84-85, 102-103, 132-133, 154-155;* Tim Hayward *– 17, 18, 26-27 (top), 56, 64-65, 88-89, 113, 123;* John Rignall *– 28-29, 42-43, 100, 136-137;* Dick Twinney *– 24 (top), 108, 140;* David Webb *– 12 (right), 24 (bottom), 26-27 (bottom), 36, 45, 54, 58, 78, 87, 106, 115, 124, 127, 144, 145, 146, 148, 152, 153.*

The Author would like to thank Carolyn Boulton, Jacques Lhonoré, Bob Stebbings and Yves Maccagno for all their comments and advice, and the Templar team for smoothly steering this book through to publication.

A TEMPLAR BOOK

Devised and produced by Templar Publishing Ltd
Kings's Head Court, Dorking

Editor: Mandy Wood
Designer: Mike Jolley
Production: Sandra Bennigsen
Typeset by Templar Type

ISBN 0·600·30574·0

Published 1985 by Hamlyn Publishing
A division of The Hamlyn Publishing Group Ltd.,
Bridge House, London Road, Twickenham,
Middlesex, England.

Printed in Spain by Cayfosa, Barcelona
Dep. Leg. B-9885-1985

CONTENTS

INTRODUCTION

All man-made habitats become colonized with wildlife. Seemingly inhospitable places like isolated oil rigs, ports, slag heaps, industrial wastelands and factories blossom with wild flowers, insects and mammals. Closer to home, too, you are likely to see a wide range of opportunist plants and animals – in city centres, suburban parks, along motorways and, of course, in your own garden.

Wildlife hasn't been slow at cashing in on man-made habitats or at learning new skills. Black-headed gulls saved themselves from extinction by becoming scavengers and moving inland. Amphibians crowded into garden ponds until they were jammed full. Opportunist sparrows learnt to make their nests entirely from industrial fibre glass. Roadside birds mopped up the hoardes of insects killed by cars, and wasps quickly learnt that peeling squashed insects from car grilles was an easy way to get fast food.

Some European wildlife is very cheeky. Snow buntings and stonechats – normally birds of the countryside – have turned to scavenging at picnic sites, a job normally reserved for magpies, rooks and crows. Other birds loiter round school playgrounds – the whistle for the end of break telling them when to drop down for spilt titbits – and black-headed gulls are tuned in to the piping sound of reversing dustbin lorries on refuse tips, a sign to them of the arrival of some more good pickings.

Towns and cities are good places for wildlife, since these environments can be up to 5°C warmer than the surrounding countryside. City plants have a few extra weeks in which to grow each year, and insects are often saved the rigours of frost altogether. Many town birds have adapted to nesting on the ledges of high-rise buildings – the man-made equivalent of their usual tree-top or cliff-side retreats – and have changed their diet to take advantage of the edible waste on offer.

After dark another set of urban wildlife takes over. Bats fly from attics and cavity walls, feral cats intermingle with urban foxes and badgers, and in some places feral deer come close to city centres. Suburban owls rely on belfries and old trees for nest sites, and city hawks hunt over the flower beds and

open spaces of parks and gardens.

If some wildlife seems to do unusual things it's because man has made suitable habitats for it quite inadvertently. Oystercatchers make 'scrape' nests out of the gravel on flat-topped municipal buildings; house crickets breed in the warmth of rubbish tips where alien cage-bird seed also flourishes; house martins build nests and successfully rear young on ferries, goldfish thrive in canals and crocodiles have been found living quite happily in city drains.

Much wildlife is present on our own doorstep without us knowing it; dippers, otters and kingfishers pass beneath our town bridges; the wildfowl of estuaries and coasts moves inland at high tide to crowd out airports, agricultural land, town parks and commons; seaside plants have moved inland and benefited from the de-icing salt accumulated at the roadside; butterflies build up large populations on the grassy slopes of motorways and parks, and motorway service stations offer a unique habitat for scavenging birds.

Some wildlife has always been encouraged by man or has shown a distinct preference for living alongside him. Robins, magpies, rooks, crows, swallows, martins and swans, for example, have all done well living in our pockets. Moreover, much of the wildlife we see on our doorstep is there because we have introduced it. We are responsible for familiar things like Oxford ragwort, fireweed, Canada geese, giant redwoods, eucalyptus, rabbits, mink, fallow deer and carp. Perhaps the most surprising thing is that wildlife has colonized every man-made habitat so thoroughly, be it along the green corridors of roads and railways, the concrete jungle of city centres or the open water of our reservoirs and gravel pits.

All these major habitats, and the flora and fauna that can be found there, are covered in this book. There are also boxed lists in each chapter which note the commonest species of European mammal, bird, insect and plant to look out for in any one place. You may be pleasantly surprised to discover just how rich an array of species there is living in your own garden already, as well as practical suggestions about things you can do to enrich it even further. Only by understanding and encouraging plants and animals – growing nectar-rich plants for example, conserving rare and endangered wild flowers or making the right type of nest box for birds or bats – can you make the most of the wildlife on your doorstep.

Everyone's garden is a potential nature reserve. There is probably plenty of wildlife there already, but with a little encouragement much more will arrive and stay. Cultivate your garden for wildlife and watch it prosper. Discovering what is there can be an absorbing hobby. Daytime visits from foxes and badgers, small mammals like hedgehogs and mice shuffling through the undergrowth under cover of darkness, a host of insects hibernating in the garden shed, and interesting wild flowers growing in the rockery are just a few of the things to look out for.

GARDENS

You can find wildlife in any type of garden, whether it be a walled garden in the centre of a city, a large rambling garden set amidst rolling green hills, or a well-maintained patch of lawn on the edge of town.

Fortunately, we all have access to gardens of one type or another – either our own or one of the increasing number of gardens or parks open to the public. But although much of the wildlife in such places is immediately apparent – from the sparrows scavenging on the lawn, to the honey-bees tirelessly visiting flower after flower – a far greater proportion of the action remains hidden from our unobservant eyes.

THE GARDEN SANCTUARY

A high percentage of the European population is made up of gardeners. Surprising then, that so many still think of the garden as a place merely to grow cultivated plants rather than as a home for a multifarious collection of living things.

All gardeners spend at least some time digging the soil, weeding, lifting stones or moving compost and, during this process of cultivation, they are continually exposed to some of the smaller denizens of the earth – the worms, insects and spiders. Moreover, you only have to glance along the herbaceous border on a summer's day to see the amazing variety of creatures that come to visit the flowers, or at the gathering round the bird table every time the breakfast scraps are put out. The question is, what else might lurk beneath the grasses or among the branches of the trees?

To begin with, you would probably be surprised by the number of wild flower species already growing in your garden. If you started with a bare patch of land you would probably find at least 200 different species of wild flower growing there after two years. The long grass may harbour all sorts of meadow flowers – buttercups, ox-eye daisies, lady's smock, stitchworts and red campion for example. A hedgerow bank may have foxgloves, vetches, bugle and even mints if there is fresh water about.

Side-by-side with the plants exist many different creatures. Resident birds come and go and migrant birds and insects may call in for short periods. You may find ladybirds sunning themselves on a well-chosen rose leaf and hoverflies flitting endlessly to and fro, while a bank vole moves silently along its well-worn passage through the undergrowth.

Gardens throughout Europe are becoming increasingly important as reservoirs for wildlife driven away from the often inhospitable countryside. Those surrounded by agricultural land, for instance, may offer the last refuge for wild flowers, insects and mammals that can no longer survive in the alien agricultural environment. Similarly, gardens adjacent to man-made conifer woodland may find themselves harbouring other visitors – from wood mice to stag beetles. You can never be sure of just who or what might be using your garden as a temporary or even permanent home. But one thing is certain, and that is you will never find out if you don't bother to look!

GARDEN TYPES

Different types of garden obviously attract different types of plants and animals. What you see in your garden someone else might never find in theirs, and vice versa. Yours might be an overgrown Victorian garden rich in trees, shrubs and impenetrable thickets. Here might be the den of a suburban fox, there a home for a shrew or other small mammal.

If you live next to an old wood you may get visits from woodpeckers, nuthatches and tree creepers. You might hear the twittering of a group of long-tailed tits as they move through the treetops in their search for food, or be afforded the welcome surprise of a charm of colourful goldfinches as they descend to feed on your thistles or other 'weeds'.

A garden full of lawns, flower beds and hedgerows might receive a visit from two or three magpies, a hedgehog on its nocturnal wanderings or an urbanised grey squirrel content to make the rounds of the birds' nut bags. Many of the colourful flowers of a typical cottage garden – lupins, marigolds, sweet peas, columbines – are very attractive to butterflies and bees. Field mice, shrews and voles may live in the thickets of stems and harvest the small insects which abound there. The hedgerow may

__Below:__ Any garden harbouring buddleia or other nectar-rich plants is almost certain to be blessed with frequent visits from butterflies – brimstones, red admirals, peacocks and small tortoiseshells to name but a few. Peacocks, like the one shown below, have evolved 'false eyes' to scare away predators such as birds and lizards. When disturbed, the butterfly will fan out its wings to show the eyes to their best effect. Like many aristocratic butterflies, the peacock lays its eggs on nettles. The resulting black caterpillars have a voracious appetite and can be found feeding on nettle clumps in wild corners of the garden.

be rich in foxgloves, buttercups and Queen Anne's lace (commonly called cow parsley), while the garden itself may harbour medically important plants such as comfrey, betony and feverfew.

Tiny courtyard gardens in very built-up areas can also support a wealth of wildlife. Small tortoiseshell and green-veined white butterflies may make passing visits in the spring, and hanging baskets or window boxes full of flowers prove an irresistible attraction for hoverflies and

***Above:** Bank voles, like the one shown here, are often found in gardens. Like their relative the field vole, they can be told apart from shrews and mice by their blunter heads, reddish fur and smaller ears. Both these voles feed on fruits, seeds and nuts and can be distinguished from each other by the length of their tails, the field vole's being longer.*

honey-bees. Such gardens also play an important part in the lives of many city-dwelling animals, as we shall see in the following chapter.

There is always more in the garden than you think, so look twice at everything. Those sparrow-like birds in the rambling roses might be warblers using your garden as a resting stop whilst on their long migration from Africa. The butterflies on the buddleia are probably more than just red admirals, and are all those white ones on the cabbages really just cabbage whites?

Take a little time to explore your own garden. Look at the different areas – the flower beds, the compost heap or the hedgerow bank – and see what you can find there. Without doubt, you will have at least one surprise in store...

GARDEN BUTTERFLIES	
Brimstone	Painted lady
Comma	Red admiral
Gatekeeper (hedge brown)	Small copper
Large white	Small skipper
Meadow brown	Small tortoiseshell

WEEDS OR RARITIES?

People often grow interesting plants in their gardens – snake's head fritillaries, pasque flowers, Welsh poppies, Cheddar pinks, spring gentians, wild gladioli, even columbines, corncockles and cornflowers – without even realising that they are wild plants now extinct in some of their natural habitats. It is often surprising to learn that what we regard as a rather common garden plant – even a 'weed' – is in fact a national rarity. So long has the break been made between the origins of the plant and the cushy confines of the garden that people are often beguiled into thinking that certain plants are truly of garden origin.

Who would believe that wild gladiolus, which is a colonist in gardens, or Cheddar pinks or lady's slipper orchids, which you can buy in garden centres, are all endangered species in some parts of Europe. In much of mainland Europe wild flowers still flourish but there are some places, Britain for

***Left:** Columbine (aquilegia) is a favourite cottage garden plant. A chalk-loving species, it is now rarely seen in its natural habitat. The flowers come in various colours, from blue to white, and the plants may be either tall or short. Once established in the garden it rarely disappears since each plant produces masses of shiny black seeds which soon grow into new plants.*

example, where there has been a serious decline in the number of wild flower species. The current estimate is that about 300 of the 2,000 British wild flower species are now seriously endangered and at least 20 species have become extinct since serious recording of them started in 1600.

Take *Daphne* as an example. Britain has two species – mezereon and spurge laurel, both natives which thrive well on chalk. Yet where else would you normally find mezereon flowering early in the year, with its terrific heavy scent, but in a garden. Similarly, Cheddar pink is another flower often found in gardens. In the wild it used to be much more abundant than it is today, but over collecting has reduced it to inaccessible crags in the Cheddar Gorge – its traditional home. Its relation, Deptford pink, was once found along hedgerows, waysides and on waste ground but it too has become a national rarity – a flower certainly not found these days in the suburbanised London borough of Deptford after which it was named. It is, however, common in other parts of Europe.

It is indeed a sad fact that today these vestiges of our past, the plants tied up in our countryside heritage, are sometimes only ever represented as plants growing in gardens.

HONEY-BEE PLANTS

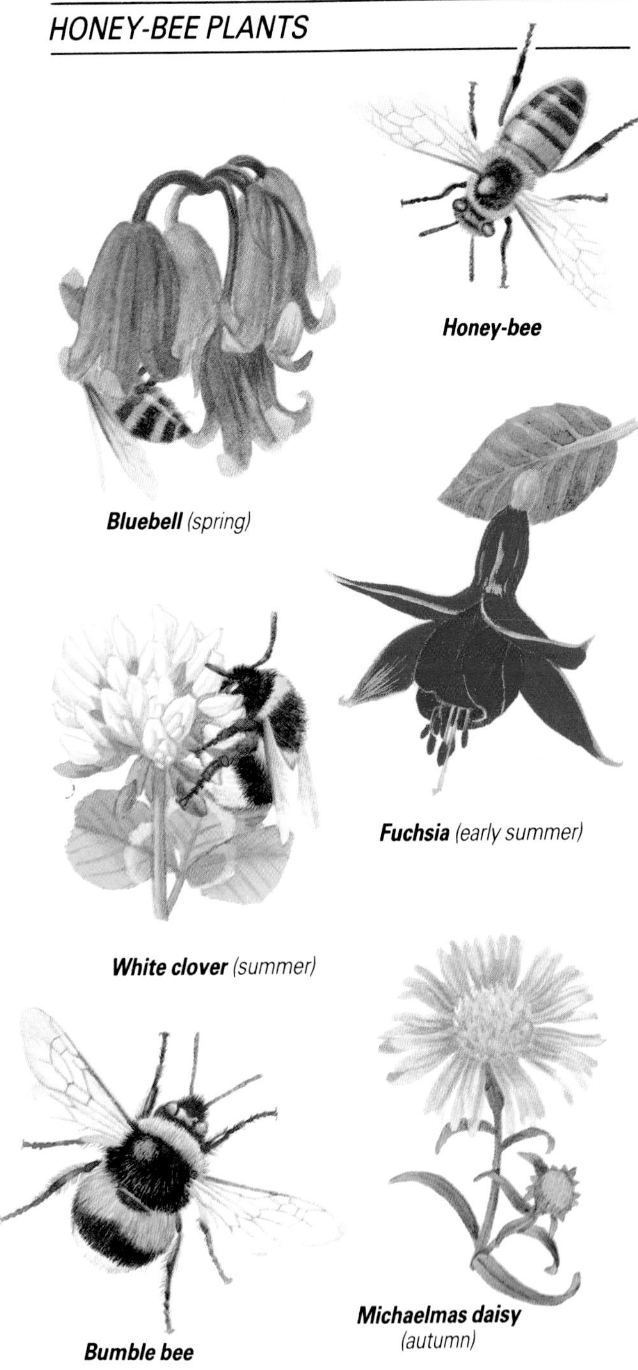

Most people will find honey-bees in their garden throughout the greater part of the year. They fly up to 3 kilometres from their hives and visit all sorts of wild and garden plants. Some flowers, however, receive more attention than others. This is because they produce richer sources of nectar or pollen. The nectar is sucked up into the bee's honey-stomach, whilst the pollen is collected in special baskets on the hind legs.

Bumble bees are much bigger insects. Early in the year you will see them roving about the garden looking for a place in which to rear their colonies. In the autumn, mated bumbles look for hibernation quarters under hedges, in banks and compost heaps. They too collect pollen and nectar and, being much heavier, cause flowers to bend with their weight.

HERBACEOUS BORDERS AND BEDS

Many gardens have herbaceous borders. The tall spikes of foxglove, mullein, delphinium and hollyhock act like magnets to many of the pollinating insects, so there is always plenty to look at.

Notice the dangling male anthers of the flowers covering the visiting honey-bees with pollen. Watch also how the bumble bees actually visit the flowers. On many plants, special petals provide them with a ready-made landing stage which depresses with the weight of the insect. This in turn brings down the stamens which give the back of the insect a liberal dusting of pollen. The bee is soon attracted to another flower where the pollen from the first plant sticks to the ripe female part of the second, thus effecting cross pollination.

Of course, this two-fold process of the bee collecting food whilst at the same time acting as a pollen-carrier is no coincidence. Wild flowers and insects evolved at about the same time and have co-evolved together. The plants produce strong scents, colourful petals and sugary nectar, all of which attract the necessary insects. Some of the latter may eat the protein-rich pollen, but a lot is transferred to other plants, thereby enabling the vital process of pollination to continue.

Further back in the border dahlias may be staked out. Earwigs love to get in amongst the petals – liking to find places where they can make surface contact on all sides of their bodies. Butterflies such as red admirals, small tortoiseshells and peacocks will be attracted to the taller asters and Michaelmas daisies, and evening moths to the tall tobacco plants and night-scented stocks. You might even see the impressive convolvulus hawk moth, though an infrequent migrant to British shores, feeding at dusk on the tubular flowers. Gardens in urban

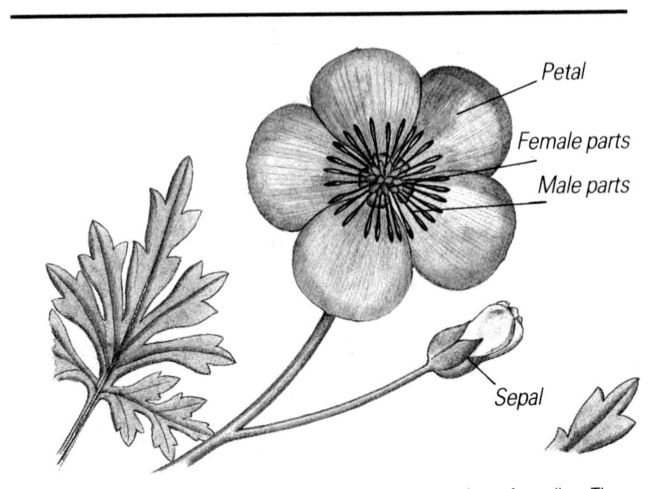

Meadow buttercups are only visited occasionally by honey-bees for pollen. The buttercup is a hermaphrodite, so both male and female parts are carried in the same flower.

Above: *Herbaceous borders attract a great many insects and other invertebrates. Shield bugs, leaf beetles, spiders and earwigs can all be found living in the leaf litter or amongst the flowers. Many eat nectar or pollen, and the pest species eat the leaves or drink the sap. When planting a border, the skill is to develop a colourful display which continues throughout much of the year, providing a ready supply of necessary food for insects such as butterflies and bees.*

areas are likely to have the privet, poplar, lime and eyed hawk moths also. The moths of this family are mostly named after their caterpillars' main food plant, apart from the eyed hawk whose caterpillars feed on fireweed (another name for willowherb).

The plants along the edge of the border have their own insect audience. Ice-plants, so named because their leaves feel cold even in the height of summer, are excellent for attracting butterflies, as are daisies, shrubby garden ragworts and primulas. The small tortoiseshell is the commonest butterfly to be found in this part of the garden, sucking up nectar from Michaelmas daisies, sweet Williams and other border plants. But if you want to find its caterpillars you'll have to look elsewhere. Like many of its relatives, this butterfly chooses to lay its eggs on quite a different plant altogether – the stinging nettle. Nettle patches attract the egg-laying females in late spring and the eggs are laid in clusters on the undersides of the young nettle leaves. You can find the spiny dark caterpillars

GARDEN MOTHS	
Angle shades	Garden tiger
Blair's shoulder knot	Heart and dart
Elephant hawk	Puss
Garden carpet	Swallow prominent
Garden dart	Yellow underwing

THE CHANGING FACE OF THE POPPY

There are many species of poppy in Europe but three of the native ones are familiar sights – the common red or corn poppy, the Welsh poppy and the yellow-horned poppy, a near relation found on sandy coastlines. Of these, the corn poppy is probably the best known.

Poppies are well-known as colonizers of waste or newly disturbed ground – one reason being that their seeds can survive deep in the earth for many years, at least decades and perhaps centuries. They are also prolific seed producers, with each plant bearing on average around 17,000 seeds. Indeed, just 1 acre of arable land could contain 113 million dormant poppy seeds all waiting to be brought to the surface so that germination can begin. Poppies are vividly remembered springing from the disturbed ground

after Wellington's Battle of Waterloo and from Flanders' fields after the First World War. And today we are still occasionally greeted with great displays of scarlet on newly dug motorway embankments.

The corn poppy is much more a symbol of the countryside, as Monet painted it, than of the garden. Yet over the last forty years it has been difficult to ignore its slow demise from its native habitats. The days of golden corn fields scarlet with its familiar blooms are all but over, at least in Britain, though they are still a common enough sight in parts of rural France. Like so many of our native plants, the poppy has found perhaps its last refuge in the garden.

Many wild varieties of the poppy from all over the world can be found growing in the border or rockery. Opium poppies, from Greece and the Orient, frequently provide tall displays of magnificent pink flowers. The milky juice of the unripe fruit contains a strong narcotic but the ripe seeds are harmless and are often used in cooking and for decorating bread. In the rockery other introduced poppy varieties may be found – alpine, Iceland and Welsh poppy (which is much more common in gardens than in its natural habitat of rocky sites in Wales and the West Country of Britain). Another favourite, often sold as packeted seed, is Californian poppy with its bright displays of orangey yellow flowers.

throughout the summer – they are sometimes so numerous that their host plants appear to have turned from green to black overnight.

Of the many groups of insect found along the edge of the border perhaps the hoverflies are one of the most obvious. These are fast-moving flies which bear a close resemblance to some of the more dangerous members of the insect world – the bees and wasps. Hoverflies are either drab in colour, and therefore easily mistaken for honeybees, or colourfully banded in yellow and black to mimic wasps. Their disguise is further improved by the fact that they are the same size as the models they imitate and also share the same body patterns. Thus they effectively con birds and other enemies into thinking that they have a sting in their 'tail', although with the hoverflies this is certainly not the case. They are quite harmless insects, spending most of their time hovering above the border or feeding from the flowers. Many of their larvae are useful to have in the garden since they eat aphids.

Above: *Hoverflies are true flies in wasp and bee colours. They mimic poisonous insects with their colour, shape and pattern but are harmless and do not carry a sting. Many lay their eggs amongst greenfly on which the larvae feed. Others act as scavengers in the nests of bees, ants and wasps. The adults are likely to be found visiting garden flowers for nectar and pollen, the males frequently hovering above the plants and making a familiar humming sound whilst doing so. This one is visiting a scabious plant.*

Look out, too, for leaf beetles – tiny insects with irridescent wing cases – or leaf hoppers decked out in brilliant stripes. They can be found on rhododendron leaves. Shield bugs are also quite widespread amongst damp, lush vegetation and can be easily identified by the horny shield behind their heads from which they get their name.

Spurges, or euphorbias, have their own story to tell. They belong to one of the largest plant families in the world and grow wild in many European woods, clearings and along waysides. They are also grown in the herbaceous border. Not many insects eat the spurges since their white latex is laced with poisonous alkaloids. This 'milk' is one of their main characteristics and was once used in some Mediterranean countries to stun fish in still pools. Conversely, the spurge hawk moth – a common European species which is an infrequent migrant to Britain – takes pains to lay its eggs on the leaves. The brightly-coloured red, black and white caterpillars are not killed by the poisons but incorporate them into their tissues so that they too become poisonous – a good deterrent against hungry birds.

***Left:** The daisy-like flowers of feverfew are a common sight in gardens and on waste ground throughout Europe. This wild member of the chrysanthemum genus has received much attention recently due to its reported medicinal properties said to alleviate the symptoms of migraine and arthritis.*

***Right:** Honey-bees visit many sorts of flowers in the garden, acting as efficient pollinators. When feeding from pendulous flowers like the comfrey shown here, the bees will push themselves deep into the base of the petals in search of nectar, often becoming covered with pollen at the same time.*

The edge of the border may also contain some wild plants that have either become adopted by the gardener as cultivars, or have sprung up as weeds unchecked by hand or hoe. Comfrey is one good example. It is a medically important plant that has been used by man for at least 2,000 years and its multi-coloured flowers are a familiar sight along riverbanks and roadsides. The large hairy leaves are not eaten by many insects, but they do provide an ideal hunting platform for spiders which lie in wait for the occasional aphid to come crawling past. The leaves also have other uses. They can be boiled down to make a sort of spinach, accredited with nutritious qualities, and Russian comfrey (sometimes known as the compost plant) can be harvested several times each year and left to rot on the compost heap. The minerals and fibres released by several plants will provide sufficient compost for an average garden.

Comfrey flowers come in a variety of hues, from white through pink and red to purple and blue – rather like the colours found in milkwort. They are tubular in shape and enormously attractive to honey-, bumble and solitary bees. The bumbles often punch a hole in the base of the flower to get at the nectar. The honey-bees which visit the same flowers then find a quicker way of getting to the sugary secretions.

HEDGEROWS

Berries and buds in the garden offer a valuable food source for birds, and nowhere are they more plentiful than in the hedgerow. The garden's edge may be drenched with all sorts of delights but the birds have to compete with both insects and fungi, both of which will rapidly destroy the fruit.

Some fruits mature very quickly and are consumed voraciously by the birds. Others take months to ripen and are seemingly left on the bushes uneaten. The elderberry is one fruit that never lasts long. Within a week the berries have either been gobbled up by blackbirds or gathered by wine-making enthusiasts.

If you have a hawthorn hedge, the fruits will be lasting and plentiful. They are produced in such enormous quantities on all the hawthorn varieties and species – white, red, double-blossomed pink, common and Midland – that the hedgerows sometimes appear red against the winter sky. The berry-eating birds, like thrushes and fieldfares, are usually too busy mopping up the spent grain in the fields to bother about the hedgerow fruits in the autumn. But in the New Year food becomes scarce and you will soon see the flocks of woodpigeons, blackbirds and others descending to feed.

Above: *Hedgerows offer a suitable nesting site for many birds, including the blackbird, dunnock, robin and song thrush. The parent birds of the young song thrushes shown here build a cup-like nest lined with mud and grass in which to lay their four or five greenish-blue eggs. The young hatch after 13 or 14 days and are fed by both parents for a similar period. The birds may use the same nest to raise another brood later in the year. You may hear the parent birds singing loudly to proclaim their territory.*

Right: *The unmistakable robin is a familiar sight in many gardens, but its drab young are often overlooked. Despite their affinity towards man, robins are fiercely territorial towards their own kind, fighting off intruders and singing loudly to establish their territories.*

Your garden hedge may well contain sloe or blackthorn – a constituent part of most hedgerows. Considering that it flowers early in the year – a blackthorn winter is in February or March – it is remarkably 'slow' at producing its mature fruits which do not appear until October or November. A relative of the sloe is the damson which is often found around allotments and in older gardens. It has larger fruits which, if left to fall to the ground and rot, are also eaten by blackbirds. The pale blue bloom on the surface of these and other similar fruits is a naturally occuring fungus which is not harmful to man – indeed it contains the yeasts which are so gainfully employed in fermentation!

In the spring, the hedgerow provides an ideal nesting sight for many different kinds of bird, particularly the dunnock (hedge sparrow), blackbird, thrush and robin. Of them all, perhaps the robin is best known to the gardener, certainly the only one bold enough to accompany him on a morning's dig in the hope of a worm or two. Robins frequently nest in thick hedges, sometimes building their nest on the ground using old leaves. This is mostly the hen bird's job, the male usually being too busy defending his territory to think of much else! Robins have also been found nesting in jam jars, boots, saddle bags and letter boxes, proof indeed of their adaptability to man and his environment.

The common practice of cutting hedges back during the spring and summer often has a

Above: *The lords-and-ladies plant (cuckoopint) often grows under hedgerows and in suburban woods and parks. The curious shape of the flower with its cowl and central rod serves a specific purpose – to aid pollination. In the spring the central rod or spadix heats up – just a few degrees higher than that of the surrounding vegetation. This attracts lots of flies which enter the cowl and become trapped by a ring of hairs surrounding the stem of the central rod. Once inside the basal chamber, the flies become covered in pollen and can only escape when the female part of the flower has been pollinated. The restraining hairs then break down, allowing the flies a free exit, only to be lured into a similar trap by another member of the species.*

disastrous effect on nesting birds. Too vigorous a cut exposes them to predation from magpies and jays, not to mention the prying eyes of humans. Unfortunately, many farmland hedgerows have been destroyed altogether – perhaps as much as 200,000 kilometres since the Second World War – hence the importance of conserving those that remain. A well established hedgerow (meaning one that has been around for several centuries) can support a great diversity of animal and plant life. Indeed, in some areas where there is little woodland, certain species of bird would be absent altogether if it weren't for the shelter provided by the remaining hedgerows. So if you have one bordering your garden, guard it well. Prune in late autumn when the birds have finished nesting.

BUTTERFLY FOOD PLANTS

Gatekeeper on privet. *Privet blossom is strongly scented and attracts butterflies. Here a gatekeeper is feeding on the nectar. Honey from privet is often very strongly flavoured.*

Small white on radish. *Leaving vegetables to bolt often results in plenty of flowers for butterflies. Here a small white feeds at radish.*

Small copper on fleabane. *Fleabane is a multi-flowered plant which is a useful nectar source for several butterfly species. Here a small copper imbibes nectar.*

If you allow some of your weeds to flower in the garden, you will have many visits from butterflies. Dandelions, thistles and fleabane are all excellent in this respect. The last two have long flowering periods and repay their presence with visits from several colourful butterfly species.

Even a lawn full of daisies will attract butterflies, and privet hedges will overflow with a mass of scented flowers if left to grow. Why not plant some scabious, lavender, majoram or chives in your garden, or invest in a few cuttings of buddleia? In the herbaceous border you could put field eryngos, feverfew, marigolds and zinneas, all of which attract a multitude of insects – not only butterflies, but bumble and honey-bees as well. A more comprehensive list of nectar-rich plants which will attract butterflies is given on page 157.

CATERPILLAR FOOD PLANTS

Buff tip moth on hazel. *These caterpillars often feed in groups which can quickly defoliate small trees and bushes.*

Magpie moth on gooseberry. *These caterpillars are typical 'loopers' or 'inch-measurers', forming a characteristic loop when they walk.*

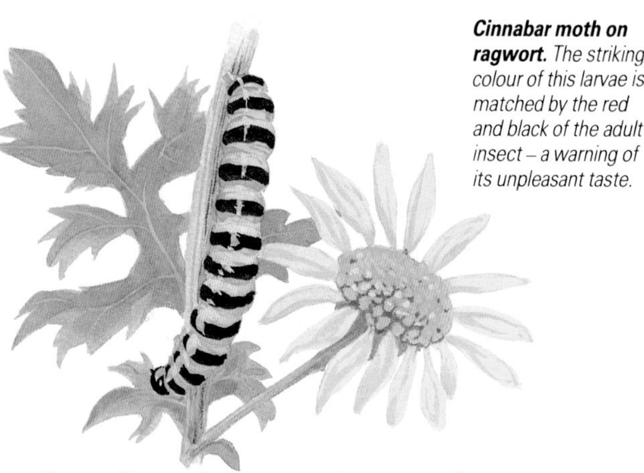

Cinnabar moth on ragwort. *The striking colour of this larvae is matched by the red and black of the adult insect – a warning of its unpleasant taste.*

Caterpillars of all sorts of insects will turn up in your garden, not just those of moths or butterflies. Look for them on your vegetables, in the foliage of your hedge or on your ornamental bushes and flowers. You might also find one crossing the lawn or path – probably a fully-fed moth caterpillar moving from the shelter of its food plant to find a suitable place in which to change into a chrysalis. Moth caterpillars are among the most colourful of larvae and usually very much brighter in appearance than the ensuing adults. Many feed in large groups, stripping their food plants of all foliage before moving on to the next plant. Others, like the hawk moth caterpillars, feed singly on such forest trees as pines, poplars and limes. Repatriate any caterpillars brought in for identification to the safety of the leaf litter underneath a hedge or bush. The hairy species are best avoided since they can cause skin rashes.

Above: *The fleshy red berries of bittersweet (woody nightshade) can often be seen around the garden. Like those of its deadly relative, they are poisonous.*

The base of the hedge, or the bank on which it stands, often figures as a great refuge for many wild plants and insects. You may find ragwort, its tall stems supporting a colony of striking black and yellow striped caterpillars soon to become fully-fledged cinnabar moths. Or wild violets – the true ancestors of the brightly-coloured pansies which so often adorn the edge of the footpath or lawn – caterpillar food plants of several species of the fritillary butterfly. Banks of periwinkle may straggle over the ground and their purple flowers are always good at arresting the brimstone butterfly for some refreshment.

With the arrival of autumn a new and often striking component of the hedgerow flora appears – the fungi. Helping to degrade the leaf litter and fallen twigs, the toadstools and mushrooms of the hedgerow make a spectacular display with their fruiting bodies. You can expect to find fly agaric – a cosmopolitan fungus with hallucinogenic powers – under silver birch trees, the green wood cup fungus – a microfungus which was once used as a green inlay on boxes – if you are close to oak trees, plus numerous other common species such as ink caps, death caps and sulphur tufts.

ORCHARDS AND LONG GRASS

Like the hedgerow, the orchard is a haven for birds – this time the bud-eaters. One of the most destructive and at the same time most attractive, is the bullfinch. You'll usually see these birds in pairs or flocks; the cock being particularly striking with his scarlet breast. Unfortunately, they are serious pests in orchards where flocks of 100 to 200 birds can descend on the trees and literally strip them bare within a few hours. The safest way to deter these and other birds is to enclose soft fruits and dwarf rootstock in a cage of plastic or rigid netting. For the bigger trees, suspended strips of bright blue and yellow plastic are often successful deterrents. At all costs, avoid the use of black cotton thread – it kills all sorts of birds that unwittingly fly into it, binding up their beaks, feet and heads.

If you're the owner of an old cottage garden you may have mistletoe growing on your apple trees. This is a semi-parasitic plant that colonizes new hosts by way of its seed which is covered with a succulent white outer layer. This is very attractive to birds. Having eaten the fruits, the birds regurgitate the seed from their crops, and a sticky covering enables some seeds to adhere to the bark. Eventually a little root will grow from the mistletoe into the main vessels of the tree and so the plant begins its semi-parasitic life, depending on its host for some of its food.

Above: *Bullfinches are most often seen in pairs and are believed to mate for life, unlike most other finches. Seeds, weeds and berries form the bulk of their diet, although their stout bills are employed to great effect during late winter when the buds of fruit trees will be taken if other food is scarce. During this time of year they can often be seen in flocks, brightening up the bare trees lining parks and gardens.*

Right: *Mistletoe is most often found on apple and pear trees but will also live on hawthorn, lime, oak, birch and a variety of other trees. In many parts of France, particularly in the north, long avenues of willow and poplar can be found covered from top to bottom in this semi-parasitic plant, often to the extent that little of the tree's own foliage is visible.*

Its name is believed to have originated from the old German word mist meaning dung, a reference to the fact that the seeds of this plant are supposed to be dispersed in birds' droppings.

THE GARDEN HABITAT

Garden wildlife thrives thanks to man. To a fox or hedgehog, the maze of fences, sheds and shrubs encloses a paradise of food and shelter. All sorts of wildlife may surprise you on your own doorstep for many species which were once at home in the countryside now find a welcome sanctuary in gardens.

Providing the right environment for plants and animals will encourage them to colonize the garden habitat even further. A corrugated sheet left lying in long grass may well provide shelter for a small mammal, a bird table will attract migrant visitors as well as common garden species and nectar-rich plants like buddleia and honeysuckle will become regular feeding places for bees and butterflies.

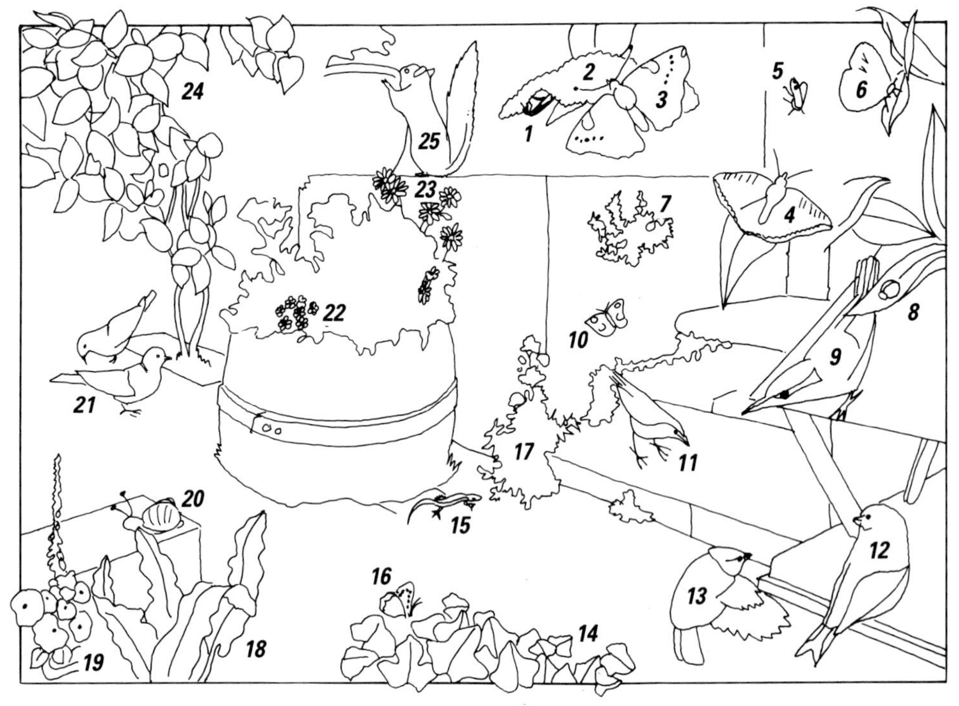

1 *Buff-tailed bumble bee*
2 *Buddleia*
3 *Red admiral butterfly*
4 *Small tortoiseshell butterfly*
5 *Honey-bee*
6 *Small white butterfly*
7 *Pellitory-of-the-wall*
8 *Two-spot ladybird*
9 *Nuthatch*
10 *Peacock butterfly*
11 *Song thrush*
12 *Greenfinch*
13 *Blue tit*
14 *Ivy*
15 *Wall lizard*
16 *Small copper butterfly*
17 *Valerian*
18 *Hart's tongue fern*
19 *Navelwort*
20 *Garden snail*
21 *Collared dove*
22 *Red oxalis*
23 *Ox-eye daisy*
24 *Honeysuckle*
25 *Grey squirrel*

The fork of an old orchard tree is a complete habitat in itself. It is often bedecked with mosses which thrust their tiny orange and green fruiting bodies or 'sporangia' upwards to release spores to the wind, or minute algae which give the bark a light green colour and are responsible for discolouring the clothes of climbers! Some trees also support little clumps of ferns, particularly one called polypody, so named because of the many rows of tiny spores on the underside of its fronds. Lichens, too, find this exposed habitat to their liking. There are many different types but the most common ones are grey or green and their fruiting bodies stand proud on the bark like mini versions of coral. Some species used to be gathered for dyeing wool, particularly one called 'crottle' which was used to dye Harris tweed various shades of brown.

Some trees may develop rot holes between their forks. Rainwater collects here and, with the help of fungi, helps to break down the wood. If the water remains there for any length of time, it may well find itself host to a batch of breeding mosquito larvae. Water beetles may also visit this mini-habitat since they fly from one water source to another at night. If you have any large

Above: *Rot holes occur in ornamental trees, large park trees, fruit trees and stumps. They often form in the flat areas between large branches and the main trunk where water accumulates and gradually rots away the inner tissues. They may also arise where branches break off. As a self-contained microhabitat they can be very rich in creatures, often ringed with mosses or ferns and alive with algae and, perhaps, mosquito larvae. Birds such as the collared dove use such water sources for drinking.*

forest trees on the edge of your garden – limes, chestnuts or oaks particularly – there may be rot holes several metres above the ground, complete with their own aquatic community.

The apples, plums, pears and other fruit falling from the orchard trees also prove very attractive to insects, providing a highly acceptable source of sugary food for many butterflies, wasps and flies. You might see red admirals, peacocks, commas, painted ladies and speckled wood butterflies congregating around the decaying windfalls, competing with other interested insects such as ants.

At the base of the tree there may be liverworts, but only if it is always damp and sheltered from the direct rays of the sun. Liverworts are flowerless plants related to mosses, but they do not grow in the rather drier places where mosses are found. The best places to find them are in damp ditches along forest trails, on the banks of streams or dykes in heathland areas or along sunken lanes. This is because they must be in a humid atmosphere at all times, preferably with the leaves being splashed regularly with water, since their reproductive cells have to swim over the surface of the liver-like lobes.

GARDEN MAMMALS	
Badger	Hedgehog
Bank vole	House mouse
Fox	Mole
Garden dormouse	Wood mouse
Grey squirrel	Yellow-necked mouse

The long grasses of the orchard provide an ideal home for small mammals – voles, shrews and mice. They thrive in these conditions of darkness and insulation, remaining relatively free from the outside world of predators and changes in temperature. If you have to cut the long grass you may become aware of small passages through the stems – the corridors through which these tiny creatures travel.

The yellow-necked field mouse is commonly found in long grass. Like its relatives the wood mouse (or the long-tailed field mouse), it is a great jumper with a long tail. The latter can be a life-saving device for, like the tails of many species of lizard, it can be shed if grabbed by a predator thereby allowing the mouse to leap to

Right: *The fleshy lobes of the liverwort reminded seventeenth century physicians of the lobed livers of man, hence its name, and were used to treat problems associated with this part of the body. Liverwort was just one of the plants mentioned in the Doctrine of Signatures, a theory proposed by Nicholas Culpeper in 1653 which stated that any plant bearing a resemblance to a part of the human body could be used to treat ailments affecting that part. Liverworts can be found in damp, shady areas and are flowerless plants related to the mosses.*

SMALL GARDEN MAMMALS

There are probably several species of small mammal already living in your garden but, because they tend to be active mainly at night and remain hidden amongst long grass or in their burrows during the day, you may see them only rarely – unless, of course, you own a cat! Look for their passages amongst the long grass in wild corners of the garden.

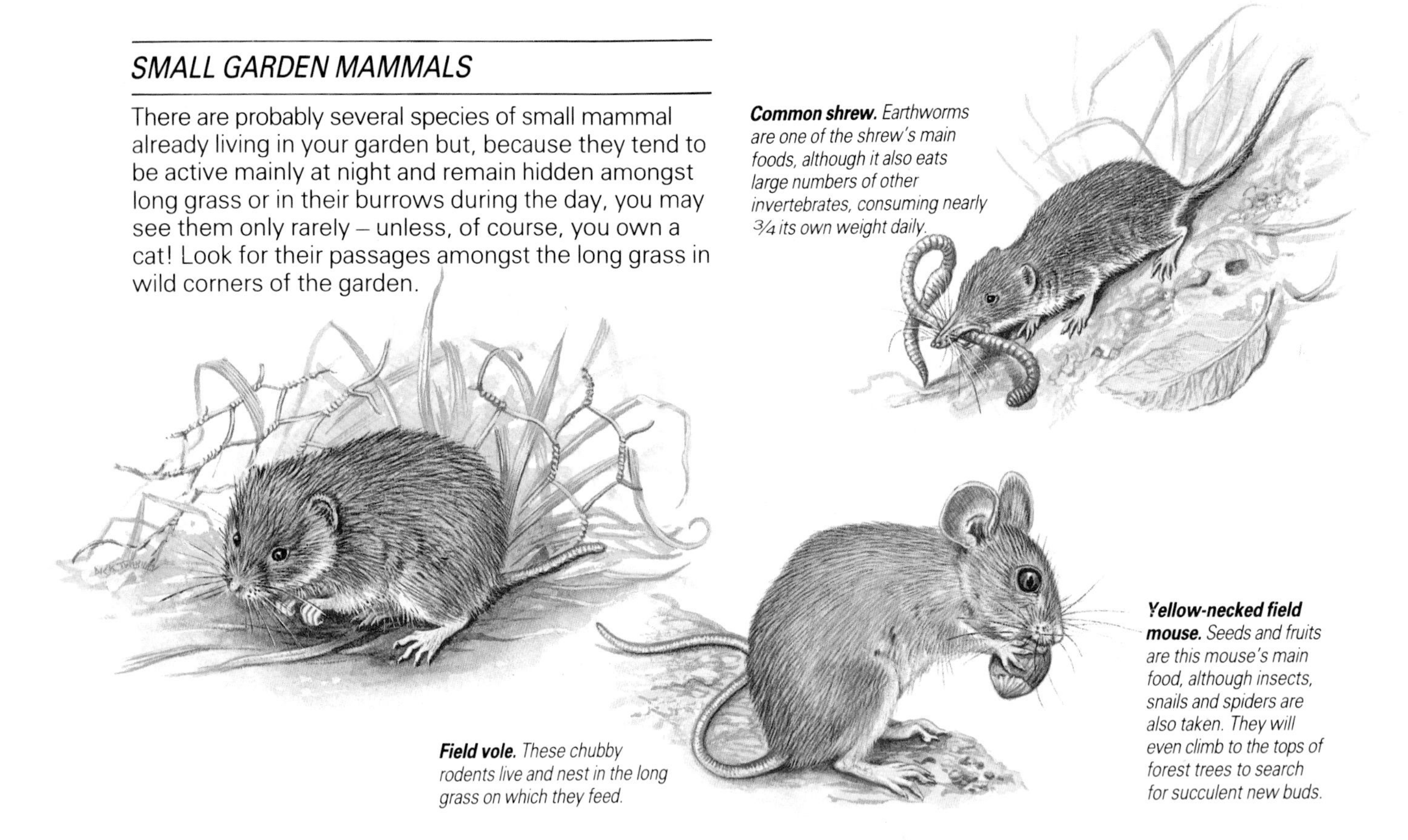

Common shrew. *Earthworms are one of the shrew's main foods, although it also eats large numbers of other invertebrates, consuming nearly ¾ its own weight daily.*

Field vole. *These chubby rodents live and nest in the long grass on which they feed.*

Yellow-necked field mouse. *Seeds and fruits are this mouse's main food, although insects, snails and spiders are also taken. They will even climb to the tops of forest trees to search for succulent new buds.*

safety. Bank and short-tailed field voles also frequent the labyrinth of grassy passages. Twenty species of vole are known in Europe, where they occupy almost every type of habitat, from mountain slopes to scrubland. They are mostly noisy creatures, often using sounds that are too high-pitched to be heard by the human ear. This is especially so during courtship or territorial fights. The vole's territory is by no means extensive, certainly no more than 1,000 square metres, whereas wood mice will frequently control an area of 2.5 hectares. Voles prefer to stay in close proximity to their nests and know every part of their home range intimately. They are at the mercy of almost every open ground predator – cats, kestrels, owls etc – and few survive for longer than a year.

Shrews, on the other hand, have relatively few enemies. Their stink glands make them unpalatable to many scavengers, although they may occasionally be killed and then abandoned by domestic cats and dogs. Their main predators are the birds of prey, particularly owls whose pellets frequently contain their bones and fur. Shrews are constantly on the move, their problem being that they have to eat a great quantity of food (over 75% of their body weight each day) to maintain their metabolism. For this reason they have to remain almost continually on the move, tracking down the next insect or worm. They only stop for very short rest periods and can starve to death in as little as four hours. The two species most likely to turn up in the garden are the common shrew and, less frequently, the pygmy shrew.

Try leaving a sheet of corrugated metal or wooden boarding on the ground – it will soon be adopted as a home by many of these small mammals as well as a varied collection of beetles, slugs, woodlice, millipedes and centipedes. You may even discover a nest belonging to a mouse or vole, perhaps with some young.

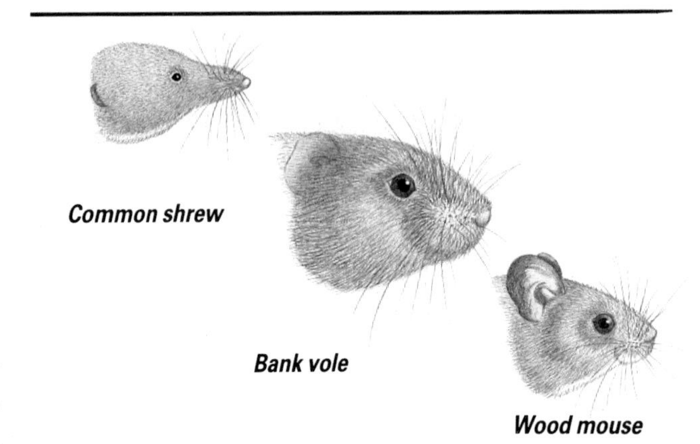

Small mammals can be placed into the three categories of mouse, vole and shrew by the shape of their heads. Voles have flatter, blunter faces compared with shrews which typically have pointed, whiskery snouts. Mice are somewhere in between. The ears are also a diagnostic feature, being larger in mice and almost completely hidden in voles.

LAWNS

Neat lawns may look aesthetically beautiful but, in truth, they are ecological disasters. Gone are the colourful spots of clover, daisies and dandelions punctuating the green of the grass. They are usually banished by the use of selective broad-leaved weedkillers, or dug up and thrown on the compost heap. Regular cutting by lawnmowers is akin to the effect of rabbits grazing. The tall, coarse grasses are replaced by finer species and the more these are cut, the more they push out – creating a rich green sward which leaves room for very little else.

Rosettes of daisy, dandelion, plantain and thistle irritate a lot of 'purist' gardeners who destroy their roots and encourage the monoculture of grass. Mosses, too, are frowned upon, yet they can give the lawn an extra splash of green, a pleasant sponginess underfoot and provide a denser habitat for many creepy crawlies that come out at night to feed.

Even if you can't bear moss in your lawn, look for it elsewhere – on tree trunks, in the rockery, on roof tops (especially bus shelters), in gutters and on sheltered banks. Thick patches of moss provide an ideal mini-habitat for small creatures such as woodlice, springtails and fully grown caterpillars which seek its sanctuary when turning into chrysalises.

Well tended lawns, however, do have some advantages. In the autumn, they can offer another welcome surprise – scores of autumn lady's tresses orchids by the square metre. This species is one of the last orchids to flower each year, preferring well managed lawns and grass tennis courts where its tiny spikes are often

Above: *The lesser celandine is found frequently in the spring in damp corners of the garden wherever the grass is left rough. Its star-like flowers and glossy leaves will reappear year after year and the bulbous swellings found on its roots have led to its alternative name of pilewort.*

Below: *Many birds can be seen in and around the garden, searching for invertebrates amongst the turf of the lawn, perching in trees and bushes or scavenging for scraps on the bird table. The robin is a common visitor that will nest in close proximity to man. These birds will utilise old kettles, saddle bags, garden sheds and garages as nesting sites and rear 5-6 young in a cup-like nest of grass and moss.*

overlooked amongst the mass of grass. Each flower spike is delicately twisted like its relative creeping lady's tresses which occurs more rarely in pine woods.

Bird visitors to lawns are numerous. Watch for blackbirds, thrushes and robins cocking their heads to listen for the slightest sound of earthworms moving through the soil beneath them. Or green woodpeckers (sometimes known as yaffles after their familiar call) probing the soil with their long bills. Groups of starlings are also a familiar sight, feeding on the invertebrates hidden amongst the roots of the grasses – particularly leatherjackets (the larvae of the daddy-long-legs or cranefly), wireworms (the larvae of the click beetle) and also ground roving beetles of various sorts. Try using a sprinkler on a fine day – you will soon see the birds descend, prepared to get wet in order to have the pickings of their prey, driven to the surface by the abundance of water.

Perhaps one of the most conspicuous visitors to the lawn is the mole. If you live in the suburbs next to the countryside you can expect about eight moles per hectare, although this figure may double during the summer. Regular forays by moles into your garden form only a small part of their extensive tunnel system which may stretch for 1,000 metres or more beneath the ground. Eliminating them, whether by traps, chemical bombs or various folklore customs, will only serve to create a vacuum into which other solitary moles will soon trundle. Moles use their tunnel system as one long trap. Small creatures fall into it only to be found and eaten by the mole doing the rounds of its circuit. Earthworms comprise the bulk of the mole's diet, the rest being made up of slugs, beetles and soil-dwelling insect larvae.

Like shrews, moles need to consume a considerable amount of food each day, a problem which they overcome in times of shortage by keeping a food store. This consists mainly of live worms which the mole immobilises by biting off their heads!

Adult moles rarely venture above ground, being badly equipped to deal with their long list of enemies (which is primarily headed by

MINIBEASTS OF THE LAWN

Minibeasts is a general term for small invertebrates such as the leatherjacket (crane-fly larva), woodlouse, centipede and springtail shown here. A whole host of minibeasts may be found hiding amongst the turf of the neatest lawn and can be easily caught with the aid of a simple pitfall trap (see page 148), and studied with the use of a hand lens. Woodlice are often very numerous and may also be found under rocks and stones, and behind rotting bark – indeed anywhere suitably damp and dark.

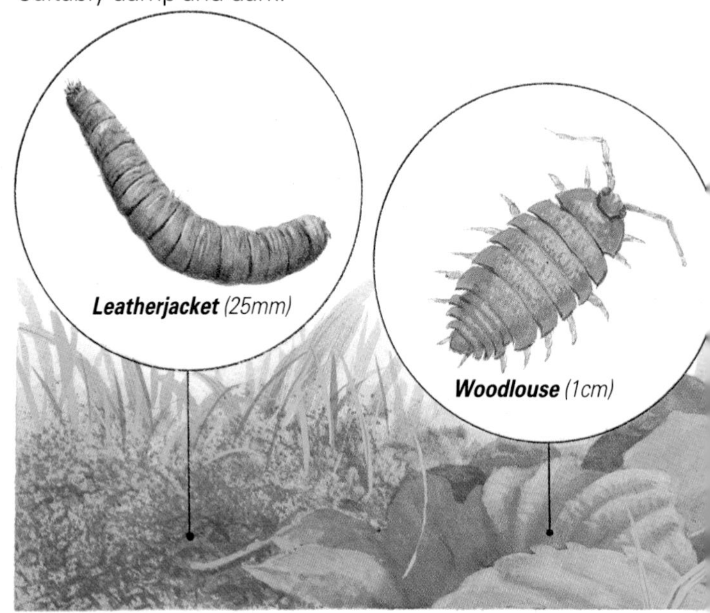

MOLES AND THEIR TERRITORY

The mole's subterranean tunnels act as an elaborate food-gathering system. Excavated a foot or so below the ground, small creatures such as earthworms and beetle larvae fall into the tunnel and provide easy victims for the mole on its next walkabout. The system will be patrolled and defended vigorously against other moles and intruders. Soil from new excavations is thrown up as familiar molehills.

Occasionally a 'fortress' may be seen – a larger-than-average hill which is used to cover a breeding nest or as a refuge during flooding. The nest itself is about as big as a football and built of leaves and grass pulled down from the ground surface. Other moles make a nest in an underground chamber and do not make a fortress.

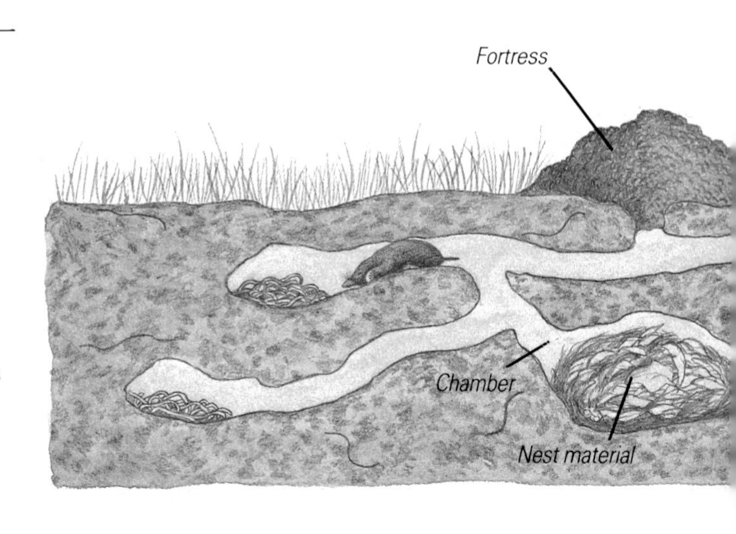

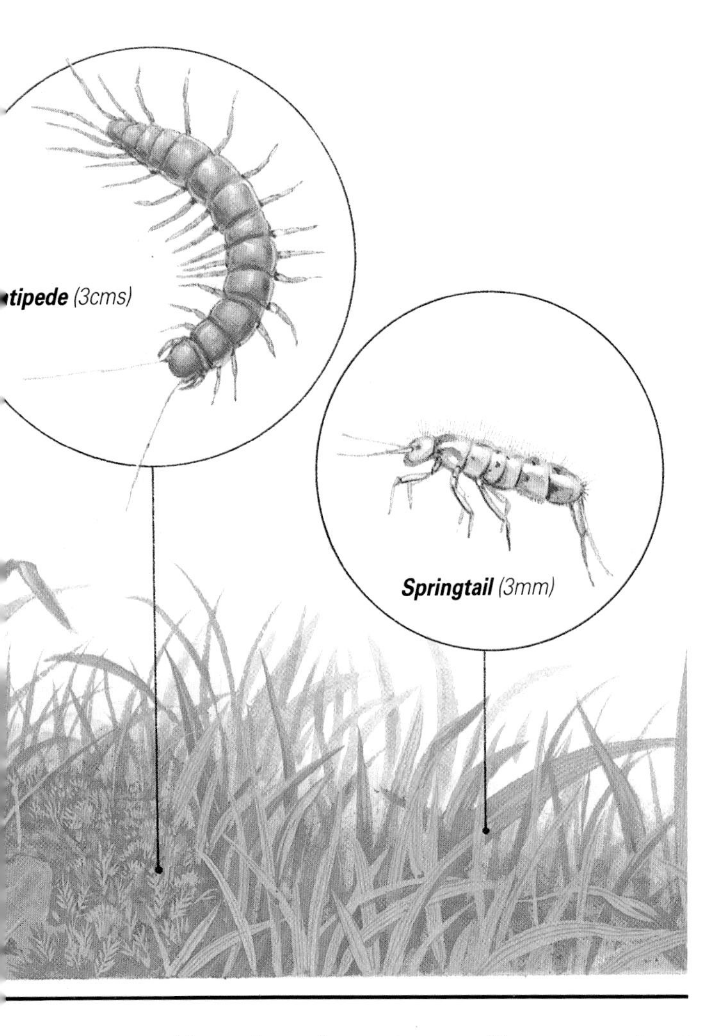

man). When they do, it is usually in search of food or nesting material, although they are occasionally forced above ground by flooding. They are purported to be good swimmers, however, and have been seen determinedly paddling towards dry land after heavy rains have caused their normal habitat to become waterlogged.

The mole's large front feet and strong claws are made for tunnelling, enabling it to move twice its own weight of soil (over 200g) each minute. Its velvety black fur will lie either backwards or forwards, thereby allowing it to slide back down its tunnels without the need to turn around. Sensitive whiskers on its snout and tail take the place of its eyes (which are only about as big as a pin head) in helping it to detect obstacles and also food.

THE WILD FLOWER SEED BOOM

Hand in hand with a growing public interest in wild flowers has come the availability of wild flower seed packeted for sale. Thanks to the entrepreneurial attitude of many seed packagers you can now plant your own instant meadow, re-establish an ancient hedgerow or buy a mixture of typical wild flowers or grasses to cultivate in your own garden.

Growing wild plants from seed is no more difficult than growing plants specially cultivated for the garden. The same rules apply, perhaps the most important being that you must choose plants suited to the soil type of your garden. The following lists contain typical hedgerow and meadow plants offered for sale as mixtures for different types of soil. Several of the species mentioned are now quite rare in the countryside so the increasing incidence of 'wild corners' for them in gardens could well be the first step towards their re-establishment in the wild.

Chalk and limestone
Black medick
Chicory
Common centuary
Common quaking grass
Harebell
Hare's foot clover
Lady's bedstraw
Muskmallow
Ox-eye daisy
Wild carrot

Clay
Common catsear
Common toadflax
Goat's beard
Kidney vetch
Selfheal
Restharrow
St John's Wort
Sweet vernal grass
White campion
Yellow rattle

Wetland
Devil's bit scabious
Dwarf Timothy
Hard fescue
Hemp agrimony
Meadowsweet
Meadow buttercup
Meadow cranesbill
Teasel
Tufted vetch
Wild garlic

Sandy
Common centuary
Cowslip
Dandelion
Harebell
Hound's tongue
Salad burnet
Selfheal
Soft brome grass
Viper's bugloss
Wild carrot

Woodland
Betony
Crested dog's tail
Chewings fescue
Foxglove
Hedge woundwort
Herb Robert
Ragged robin
Red campion
Valerian
Wood avens

Hedgerow
Dandelion
Feverfew
Goat's beard
Hedge woundwort
Meadowsweet
Ox-eye daisy
Red campion
Traveller's joy
White campion
Wood avens

WALLS

Ivy which completely covers a wall or makes up a large proportion of a hedgerow, is an excellent habitat for a wide variety of wildlife. Birds nest inside its thick masses and insects feed on both its leaves and flowers. Gardens in southern and central England with plenty of ivy will often be visited by the holly blue butterfly – the first and only blue butterfly to be seen early in the year, from the end of March through April and May. It is the only butterfly in Europe which has two food plants, its first generation of caterpillars feeding on holly, the second on ivy. Look for the caterpillars, which look uncannily like the ivy buds on which they feed, during August.

In the autumn, the yellow balls of ivy flowers provide a feast of pollen for honey-bees. It is, in fact, the last source of this vital protein food to be found before the onset of winter. On sunny days during October and November you might be forgiven for thinking that a swarm has taken up residence amongst this woody climber, so great is the noise produced by bees collecting pollen from the ivy flowers.

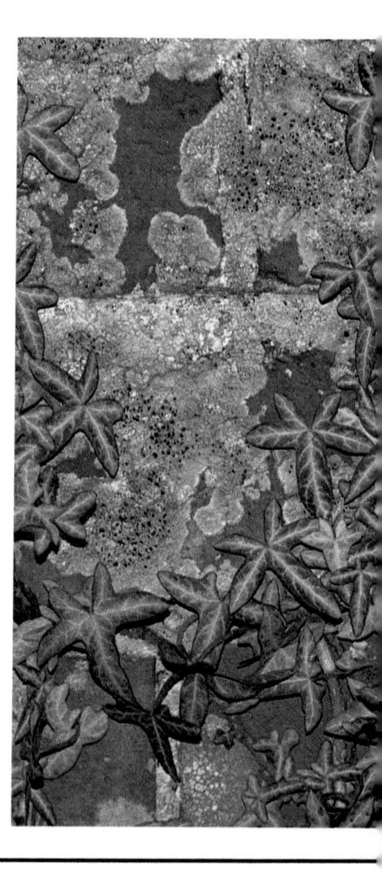

Right: *Ivy is an opportunist plant that grows in many different habitats, climbing vertically up walls and trees or creeping horizontally across the woodland floor. Its flowers produce a prolific amount of pollen – an essential supply of protein for honey-bees late in the year – and its fruits are a main component of the diet of many fledgling birds in the spring. Its leaves come in different shapes, depending on whether they are on a flowering or non-flowering stem.*

A GUIDE TO SOME TYPICAL GARDEN BIRDS

The range of birds likely to be found in your garden depends very much on your proximity to woods, parks, the sea, freshwater and other houses, and also on whether or not you put food out for them. There is no doubt that if you do feed the birds regularly then you will attract species which you would not necessarily see otherwise.

If you live in a coastal town, visits from black-headed gulls, herring gulls and feral pigeons may be commonplace, plus the occasional surprise of a migrant waxwing. In a country garden, treecreepers, nuthatches and green woodpeckers might be regular visitors. Most gardens however – whether in town or country – have not escaped the attention of the ubiquitous house sparrow or the familiar chaffinch, robin, song thrush or blackbird. Some garden visitors are not so easy to tell apart though, and can be easily confused with their similar relatives. A few examples are shown here.

GARDEN BIRDS	
Black redstart	Magpie
Bullfinch	Nuthatch
Chaffinch	Robin
Collared dove	Starling
Dunnock	Swallow
Flycatchers	Swift
Garden warbler	Blue tit
Grey wagtail	Great tit
House martin	Tree creeper
House sparrow	Wren

Dunnock. *This secretive bird, once known as the hedge sparrow, keeps to hedgerows and the thickets of bushes. Close to, it is very attractive with its grey head and chest. It comes to the bird table in cold weather but otherwise prefers to potter amongst the leaf litter looking for small insects.*

Immature robin. *Juvenile robins lack the familiar red breast of the adult birds. For this reason they are often confused with dunnocks or flycatchers. The red feathers do not appear until late summer.*

Above: *The holly blue butterfly can be seen in the garden during spring and autumn and like other members of the blue family, has distinctive black and white antennae and legs. Its chrysalises can be found on holly leaves during the spring or amongst ivy during the winter. It is a unique species of butterfly whose larvae have two different food plants according to their generation.*

Great tit. *The big brothers of blue tits, these birds can be recognised by a characteristic 'chink-chink' call, and the greater area of black on their heads.*

Chiffchaff. *This warbler can be distinguished from the willow warbler (below) by being paler above and darker below. It also prefers taller trees. Its song is often a repetitive 'chiff-chaff'.*

Willow warbler. *These birds have a distinctive song – a set of descending notes which, once heard, are said to be never forgotten. They migrate the 4,000km each year from South Africa to breed in Western Europe.*

Blue tit. *These birds are regular visitors to the bird table and will cling eagerly to nut bags to peck at the contents. They are successful breeders and may take to tit boxes almost as soon as they are put up.*

By Christmas and the New Year, the flowers will have turned into clusters of green berries, attracting a new collection of consumers – the birds. Woodpigeons and fieldfares are great gluttons in this respect. So, too, are the blackbirds when, by Easter, the berries will have turned into clusters of juicy, black fruits, eagerly sought by these birds as food for their hungry young.

As a refuge for nesting birds, the ivy remains almost unrivalled. The wren, one of Europe's smallest birds, is a specialist at living in ivy banks. The cock bird goes to the trouble of making several dome-shaped nests of moss and lichen, only one of which is selected by the female and lined with hair and feathers. If food is plentiful, however, the male may have more than one mate and produce as many as ten young during one spring. The ivy-clad wall also provides such birds with an ideal hunting ground for insects, a great many of which are devoured each day.

Apart from ivy, many other wild plants can be found colonizing old walls. One of the most attractive is ivy-leaved toadflax which cascades down in long growths, punctuated with dainty purplish flowers. The most interesting thing about this plant is its method of ensuring successful seed dispersal. During the autumn the seed head turns inwards towards the wall, thus allowing the seeds to fall into the cracks and crevices where germination is most likely. Once again, Mother Nature provides a device whereby the continuation of the species is almost assured.

Wallflower is another attractive plant often found on derelict houses, castle walls, in abandoned rockeries and other rough, rocky places. Like many of the other plants which live out their precarious existence sprouting from wall or stone, the wallflower can survive with little nourishment, its roots spreading deep into the wall in search of moisture. Wild wallflowers are naturally yellow-orange, but cultivars can be seen sporting all sorts of colours.

Other plants on walls include navelwort, pellitory, fumitory and yellow corydalis.

GARDEN WILD FLOWERS	
Bramble	Honeysuckle
Buddleia	Ice-plant
Clover	Ivy
Comfrey	Lady's smock
Crocus	Lavendar
Daisy	Marigolds
Dandelion	Michaelmas daisy
Feverfew	Nettles
Fleabane	Shepherd's purse
Foxglove	Wallflower

Left: *Wild plants and garden escapees are found frequently growing from cracks and crannies in walls – the mullein shown here is one example. The leaves of this plant are soft and hairy and are sometimes eaten by the caterpillars of the mullein moth which hide amongst them from birds and other predators. There are several species of mullein in Europe and they all grow tall. They are biennial and produce a basal rosette of leaves only in their first year. Apart from mullein, many other plants grow on walls – willowherb, buddleia, fig, and wall rue are other examples.*

Above: *Navelwort has succulent rounded leaves with a dimple at the centre resembling a belly-button or navel, hence its name. The flowers of some specimens are quite spectacular, rising in tall spikes from the plants' often precarious position on stone walls and rockeries.*

Navelwort has fleshy circular leaves with a belly-button like depression in their centre. In damp areas an introduced plant called mysteriously 'mind your own business' or 'mother of thousands' can form dense matts of vegetation on walls and in rockeries. Recently this plant has been exploited by the potted plant business and mass-produced plants now appear in garden centres at exorbitant prices – ironic, since the wild plant is very difficult to keep in check once established in the garden.

Lift the cascading vegetation of any plant away from the wall and, just as in the rockery, you will reveal a teeming variety of creatures hiding underneath. Garden snails often work their way

Left: *Pied wagtails are regular visitors to gardens in rural areas. They are nervous birds which can be seen flitting back and forth across the lawn in their search for insects, characteristically wagging their long tails as they do so. Large congregations can also be found in urban areas, usually on the roofs of warehouses and factories, and around farm buildings where they feed on the flies.*

deep into the cracks or between the stones during a hot summer in an attempt to escape from the harmful drying effects of the sun. There are also opportunities for colonies of ants, cracks for earwigs, centipedes and millipedes, as well as an almost unbelievable number of woodlice. With all this invertebrate life it is hardly surprising then that, in Europe at least, such habitats are a favourite haunt of the wall lizard. These creatures are great sunbathers, deriving their energy from the sun whilst at the same time waiting for their prey of small flies, grasshoppers and other invertebrates. They may also be found basking on patios and stone steps.

PONDS

Gardens with ponds open up a new dimension, both for wildlife and for the observer. There will be the opportunity to see dragonflies, damselflies, frogs, toads and newts in their various stages of growth, and the pond may receive visits from thirsty or dirty birds and possibly hungry grass snakes. In suburban and rural areas the pond may play fleeting host to a few interesting water birds. You might be treated to a flash of brilliant blue as a kingfisher dives for its supper, or the sight of a stately heron wading through the water at dusk. And lifting the stones at the edge of even a small pond might well reveal the gaily coloured belly of a male newt in full breeding condition.

If you have built your own pond, or have one that is newly established, it can be interesting to watch the colonization of the water by pondweeds and water lilies, not to mention the rapid spread of duckweed and algae. These will be your first colonists, soon to be followed by a multitude of aquatic insects, living both in and on the water. Pond skaters, held up by the water's tension, comb the surface for unfortunate victims such as caterpillars and aphids which fall onto the water and disturb the surface film – an advertisement of their presence which the skater is bound not to ignore. Beneath the surface lurk the larvae of many aquatic insects – wriggling larvae soon to grow into midges,

***Above:** A great deal of enjoyment can be had from a simple garden pond. Water plants can be introduced and their growth followed, although some degree of control may be necessary if they become invasive. During the summer the pond may be colonized by dragonflies and other aquatic insects, including diving beetles, water boatmen and the ubiquitous pond skaters.*

If you wish to make your own pond, it need not be too expensive. First you must excavate a suitable hollow, clear it of any sharp stones or sticks, then lay a tough sheet of plastic. Bed the roots of any water plants in a soil clump and then fill the hole with water. Alternatively, you can buy a ready-made pool and sink it into the ground. Your pond can be stocked with frog spawn introduced from one which has ample and newts may arrive by themselves. If you stock fish then make sure that a passing heron doesn't get them – put plenty of vegetation around the edge, steep sides and, if persistent, wire netting over the top.

voracious nymphs soon to turn into dragonflies -- as well as a multitude of water beetles and bugs. With the help of a small net and a magnifying glass, the contents of a pond can easily provide an afternoon's entertainment for both adults and children.

It is a common misconception that frogs and toads spend their entire life in or near water. In fact, they simply breed there – frogs laying their masses of spawn early in the year followed by the strings of toad spawn in the spring. Both creatures instinctively travel to water at the start of the breeding season – often to the same pond, ditch or lake that they themselves were born in. They have been known to travel up to two miles in their search for water, often crossing motorways and other man-made perils in their travels. Once mating has taken place, the female lays between 1,000 and 4,000 eggs which the male then covers in sperm. The tadpoles soon hatch from the fertilized eggs and begin to eat algae, later moving on to flies and beetles. The transformation – or metamorphosis – from egg through tadpole to frog usually takes about ten weeks.

Toads and frogs are relatively easy to distinguish, frogs having a more athletic frame and smoother skin. Toads are more robust and warty in appearance. Moreover, they have no need to jump out of the way of predators as frogs do. Instead, the toad simply sits tight, relying on the defensive secretion that exudes from its tough skin. This poison is strong enough to protect the toad from inquisitive enemies such as cats and dogs but will not deter a hungry grass snake who simply swallows the toad whole!

Common and marsh frogs are most frequently found around garden ponds, although on the continent edible and tree frogs may also be found spawning in ponds and wells. Of the two European species of toad, the common toad is the only one likely to be found in gardens where it can frequent drier areas unpopular with frogs. The natterjack toad, famed for its resonant breeding croak, is now a protected species confined to a few sandy areas in Britain.

Above: *Despite their large size, toads are easily swallowed whole by grass snakes. Torrential summer rain brings out these amphibians from their damp retreats under rocks. Grass snakes have no poisonous venom but rely on the loss of blood brought about by wounds inflicted by their sharp fangs to immobilise their prey. After half an hour or more of fruitless struggling, the toad falls back exhausted. Its poisonous skin is no deterrent to the hungry snake.*

Below: *Inside a nest of grass and leaves, the bumble bee begins its summer work of cell building. Unlike the honey-bee, the cells of the bumble are rounded, arranged in irregular clumps and far fewer in number. The large queen straddles her developing young to incubate them – even though she is cold-blooded. Note the honey glistening in the special honey-pot.*

COMPOST HEAPS

The gardener's compost heap offers an excellent habitat for many creatures. Its warm interior is a retreat for numerous types of worm, provides a further nesting site for small mammals, snakes and bumble bees, as well as a germinating hot bed for wild flowers.

The compost heap simulates, in part, the natural decomposition of the woodland floor. One cubic centimetre of soil in an oakwood may contain six to ten million bacteria and one to two kilometres of fungal filaments. In a square metre of soil there may be a thousand different species of animals – 300 to 500 nematodes and worms, half a million mites and springtails, and up to ten thousand other invertebrates. Most of them are microscopic and their importance lies in the fact that they break down the organic matter of plants – a necessary job which they do 80-90% efficiently.

The grass cuttings, prunings and old vegetable scraps cast onto the compost heap are rich in organic matter which is eventually broken down into its constituent minerals. During the process of decay, the rotting material attracts many types of garden fly. Their eggs are frequently laid amongst the compost and, once hatched, the larvae help to break the decaying mass down still further, ready for action by fungi. Owl midges, soldier flies and fruit flies are all commonly found around compost heaps.

Digging in the compost heap may reveal the delicate nest of a bumble bee. These insects choose the drier areas where there is an abundance of hay, old leaves and twigs. They also commonly adopt the abandoned nests of mice and voles in hedgerow banks. Unlike the honey-bees, the bumble bee community does not survive through the winter. The males die off having mated at the end of the summer, and the nest is left abandoned as the females set off in their search for a safe place to hibernate. Should you find such a nest in your compost heap, look at the cell arrangement of the comb. The bumble bee's cells are rounded and arranged in irregular clumps within the delicate shell of grass, moss or leaves – not nearly so organised as the orderly nest of the honey-bee. Female bumble bees incubate their brood in early summer by covering it with their bodies and outstreched legs – quite a feat for a cold-blooded insect.

Hedgehogs, too, use the drier areas of compost heaps in which to nest and hibernate. You might also find them under your garden shed, amongst piles of autumn leaves, behind log piles or under thick grass or shrubs. They are basically solitary mammals, pairing up for only a few weeks in the year in order to mate, and have made adaptations to urban life with some success. Tracking hedgehogs with radio transmitters has indicated that they will forage through at least half a dozen gardens on a typical night in suburbia, mopping up many pest species along the way – slugs, beetles, caterpillars, leatherjackets and snails to name but a few. They will also take milk, fruit and other scraps if these are available and have been known to deliberately knock over milk bottles to get at the dregs.

Grass snakes sometimes choose compost heaps or muck heaps on farms as prime egg-laying sites. The heat generated from the decaying matter is useful for incubation and occasionally several females will choose the same place to lay all their eggs – obviously paying little heed to the warnings of the old adage. The advantage for the grass snake is that successful hatching is most likely in such a place; the disadvantage, that if a fox or rat or other predator comes across the eggs they might all be destroyed in one fell swoop. However, one

GARDEN FUNGI

Birch polypore	Horse mushroom
Cramp balls	Orange peel
Earthball	Parasol
Fly agaric	Puff ball
Honey fungus	Shaggy ink cap

CONFUSING WILD FLOWERS

The study of wild flowers provides a challenge to naturalists and there is no better place to start studying them than in the garden. If you set out to record all the different types of wild plant, you may find that after a year or so you have listed over 100 species. If you are moving soil or undergoing some excavation, study the plants which spring up on the disturbed ground. They may be long lost species which were grown in your garden over 100 years ago, for the seeds of some wild flowers will lie dormant in the soil for this length of time, waiting to be brought close enough to the surface for germination to begin.

In attempting to distinguish similar plants, pay attention to the leaf and petal shapes, flowering time and habitat. All can act as keys to identification.

Herb Robert. *A familiar wild plant, this may grow in the rockery or on an old wall. It has elongated fruits and was once used as a herb for treating a variety of ailments. It belongs to the same family as the cut-leaved cranesbill shown opposite but can be told apart by its larger leaves and five rounded petals.*

Cut-leaved cranesbill. *There are many different species of cranesbill – or wild geranium – several of which occur on waste ground or the margins of arable land. This species can be recognised by its finely divided leaves and tiny, heart-shaped pink flowers.*

Dwarf mallow. *Your garden may support one or two species of mallow, including this dwarf species. They all have tough roots, hairy leaves and flowers which are invariably mauve or purply-pink. They attract honey-bees and butterflies and are generally taller, stockier plants than the cranesbills.*

Left: *A widespread European mammal, the hedgehog's hairs have evolved into defensive spines. It is a frequent visitor to the garden where it serves a useful role as an eater of pests such as slugs, snails and beetles.*

Below: *Fruit flies are a familiar sight in house and garden. They breed in rotting fruit and other decaying vegetable matter and are particularly common around compost heaps and dustbins.*

such egg laying site was found to have a record 1,500 eggs, so clearly the advantages must outweigh the disadvantages.

Adders can also be numerous in rural gardens with warm aspects. These snakes, like lizards, will come out to bask in the sun. Slow-worms – legless lizards – are another familiar sight in gardens where there is plenty of long grass.

SEX, APHID STYLE

The aphid is probably the most famous, or infamous, of all garden pests. And this has much to do with the fact that it can reproduce at an astonishing rate – from one female over a million descendants can be produced in the space of one short summer! Aphids have perfected their methods of reproduction in response to seasonal changes in their food supply. In order to make the most of the tender young shoots of plants which appear in the spring and summer, the aphids have taken a short cut to rapid reproduction – virgin birth or parthenogenesis. By cutting out the need to find a male partner, and therefore all the rigmarole of normal sexual reproduction, the pregnant female aphids can

Ants have learnt to make the most of the ubiquitous aphid, feeding on the excess plant sap, known as honey-dew, passed out by these insects. During the autumn the ants will even take aphids down into their colonies for protection over the cold winter months. The aphids get free accomodation and food; the ants get a ready supply of their sweet honey-dew and will even stimulate the aphids to produce it by stroking them. In the spring the aphids are carried back up onto tender new plant stems where they continue to prosper, the spring sap stimulating growth. Wherever aphid colonies build up, ants can usually be found factory-farming them in this way.

continue to give birth to their young for some time. This is one reason why prize rose bushes, lupins and many plants of the kitchen garden seemingly become covered overnight in a heavy infestation of wingless aphids. Here 'wingless' is a key word, for if such virgin births continued throughout the year the strength of the aphid population would gradually decline through inbreeding. So in the autumn, the females produce winged sexual forms which are able to fly off to new localities, mate and set up more hardy colonies.

There are many different species of aphid, including the well-known greenfly and blackfly and other species which build up large populations on plants. Some aphids, such as the spruce pineapple gall aphid, cause galls to form on their hosts plants, rather like the oak apples caused by gall wasps.

KITCHEN GARDENS

Just as in our flower gardens, the kitchen garden is made up of many plants which originated in the wild. Some plants can still be found growing in their natural habitats, but many have changed so dramatically in appearance that you would not recognise the plant growing by the wayside as the ancestor of the tasty vegetable on your dinner plate!

Much of the produce from the kitchen garden still adorns the meadow, cliff or shingle beach in its natural state. For instance, it is only some 2,000 years since we took the wild cabbage from its cliff retreat and placed it firmly in the kitchen garden. From it, we selected plants with big buds to develop the Brussel sprout, took those with fat hearts and developed the cauliflower, or went for tall stems and got the kale now commonly fed to cattle. Carrot, parsnip, celery, and angelica are all typical European species and can be found in their wild state quite readily, along with introduced plants such as wild asparagus (from the Mediterranean) and potato. In most cases the roots of these plants will not be as swollen as those of their cultivated relatives, nor their fruits so big. But nevertheless they still have the genes that have enabled us to exploit their potential as a food source.

Perhaps because of the intensive cultivation of such plants, it is in the kitchen garden that the war against insects – one of the most successful animal groups on earth – continues unabated. Constant battle is waged against slugs, snails and numerous other unfortunate creatures, but it is undoubtedly the aphids that have earned our greatest disapproval. These common garden pests drink the sap from young, tender shoots using their fine syringe-like mouthparts, inflicting damage to the plants and at the same time transmitting disease-carrying viruses and bacteria. Thankfully, the kitchen garden also harbours insects that are beneficial to the gardener – ladybirds and their larvae that can chew their way through several hundred aphids each day, or ground roving beetles which come out at night to mop up a large number of small beetle pests, for example.

However, the predations of such insects are not enough to stop the aphid population explosion of the summer months – for aphids are specialists at virgin birth. A mother, without any

THE LIFE-CYCLE OF THE SMALL TORTOISESHELL

Butterflies, moths and beetles go through four stages in their metamorphosis or change in form. Eggs (ova) are laid by the adult female – in the case of the small tortoiseshell, in groups on the leaves of stinging nettles. These hatch into caterpillars (larvae), which change into chrysalises (pupae) which finally change into adults (imagines). Some insects go through this complete metamorphosis once a year; others repeat the process through two or three generations. With the small tortoiseshell, there are two generations of butterflies each year – the first appears in June/July from eggs laid in May, and lays further eggs which produce a second generation of adults in August/September.

Caterpillars are notoriously difficult to identify, changing their shape, size and colour as they get older. Some are equipped with elaborate horns, eversible structures, irritating hairs and even deterrent sprays to ward off birds and lizards. The black and yellow caterpillars of the small tortoiseshell are found in colonies, often spinning a silk web over the tips of nettle leaves.

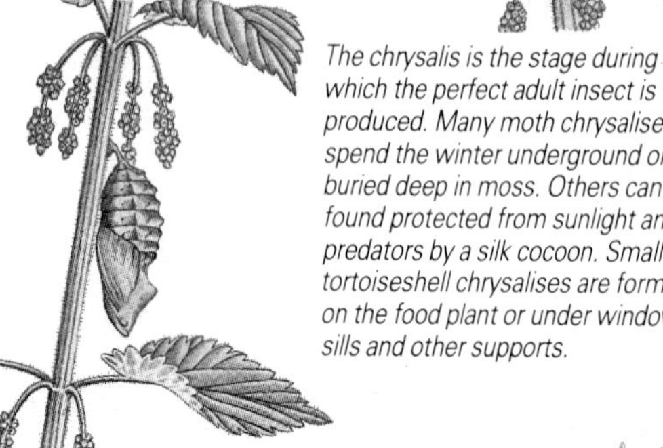

The chrysalis is the stage during which the perfect adult insect is produced. Many moth chrysalises spend the winter underground or buried deep in moss. Others can be found protected from sunlight and predators by a silk cocoon. Small tortoiseshell chrysalises are formed on the food plant or under window sills and other supports.

Eggs are fascinating things to study with a hand lens. They come in all shapes and sizes and, like those of the small tortoiseshell shown here, are often glued to the leaves of the caterpillar's food plant. Some are brightly coloured and contain deterrent chemicals passed on from the previous generation of caterpillars.

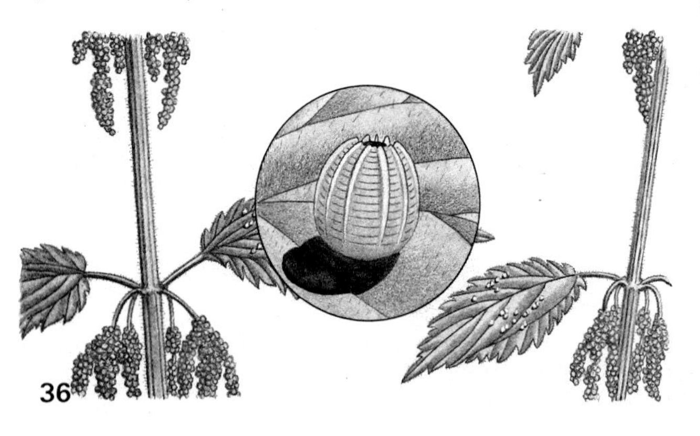

Adults are perfect insects, complete with a million tiny wing scales, compound eyes and sensory systems. Finding a daily supply of nectar is the main priority, and butterflies like the tortoiseshell may be seen in groups feeding from nectar-rich plants such as buddleias, ice-plants and thistles.

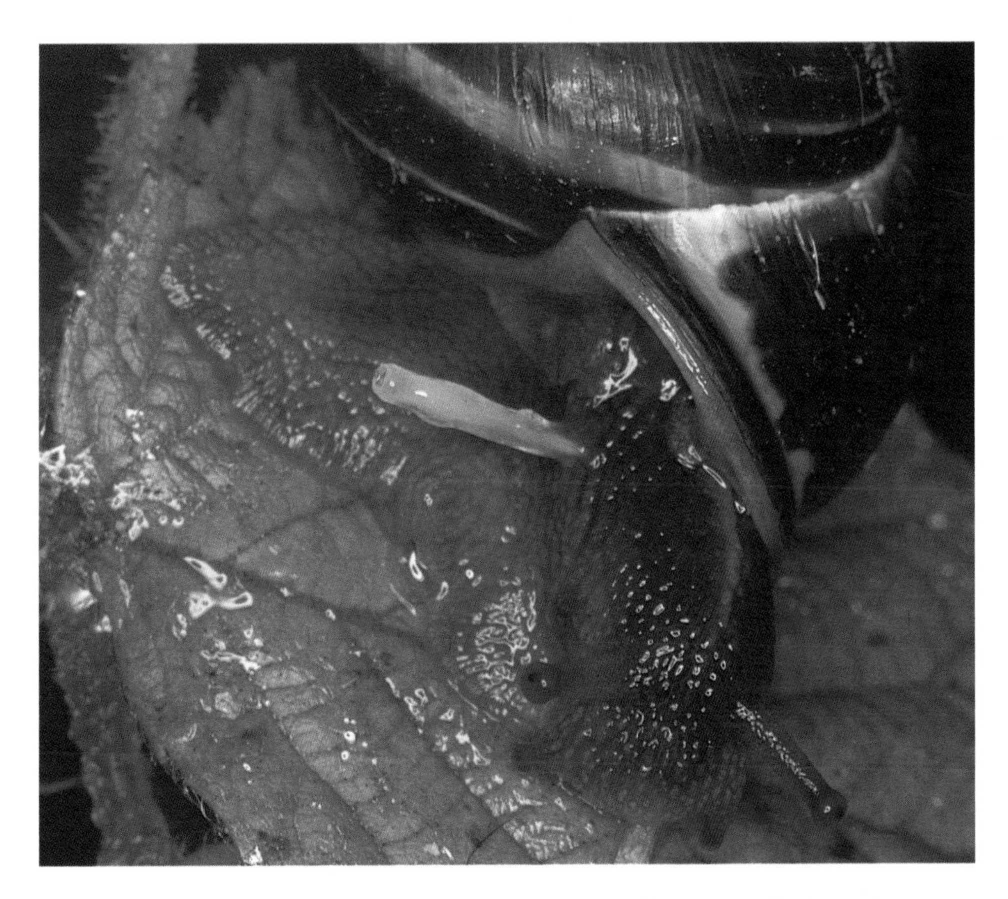

***Right:** All snails are hermaphrodites and have both male and female organs. They can often be seen clasped together in typical mating posture on wet summer evenings. Before sperm is exchanged between individuals, 'love darts' are thrown by one snail into the flesh of the other and these white harpoons are often quite visible, impaled in the snails' sides long after mating has taken place. Both partners then lay up to 40 chalky white eggs in a shallow depression in the ground. They hatch within a month.*

recourse to a male, can produce a million or so young in the space of one season! The offspring are fully formed at birth, although smaller.

The arrival of pests in the kitchen garden can be from a number of sources. Some, like aphids, free fall out of the sky, having been carried by wind currents, sometimes across continents. Others hibernate in the garden whilst still more arrive as migrants, following the same routes during spring and early summer that were taken by their ancestors. Ladybirds, for example, can migrate across the Channel, having a different energy system from man which allows them and other insects to travel greater distances.

Another noted garden pest is the ubiquitous slug. Distraught gardeners eagerly seek these invertebrate molluscs, armed with a veritable barrage of poisonous chemicals. The slugs run the risk of being sprinkled with salt, drowning in pits of old beer, succumbing to the rigours of brightly coloured slug pellets or simply being transported unceremoniously to foreign parts.

It is unfortunate that both slugs and snails cause such damage in the kitchen garden to tender young plants and seedlings. At the front of their long slimy foot, both creatures have a rasping organ – a radula with over 1,000 backward-pointing teeth. They use this to break off the surface of the plants over which they crawl. Lettuce, cucumber and marrow seedlings figure high on the slug's menu, together with the occasional foray onto lupins and clematis.

During wet summers these creatures cause havoc underground, eating out whole potatoes to leave only a hollowed-out skin or by grazing the surface of carrots and radishes.

During the winter months, snails hide away in old walls or piles of masonry, sealing themselves off from the outside world with a tough membrane which seals the entrance to their shell hole to prevent dehydration. The garden snail and striped snail are those most likely to be found in the garden, although some gardens in chalky districts are graced with the large and elegant Roman snail, thought to have been introduced by the Romans from the Mediterranean. This species also occurs on calcareous grasslands where it unfortunately falls prey to collection by restauranteurs – even though it is a protected species in several European countries.

GARDEN INVERTEBRATES	
Ants	Froghopper
Aphids	Garden slug
Birch shield bug	Garden snail
Bluebottle	Garden spider
Bumble bees	Hoverfly
Centipede	Maybug
Cranefly	Meadow grasshopper
Dark bush cricket	Silverfish
Earwig	Wireworm
Flea beetle	Woodlouse

Towns are warmer than the countryside and offer plenty of novel habitats for all forms of wildlife. Wild flowers push up in the most unexpected places – sprouting out of old walls, guttering, chimney pots and pavements. Many urban insects and mammals are notorious nuisances – cockroaches, ants, rats and mice to name but a few. Birds use town and city centres for roosting and feeding places and can often be seen in great flocks, like the starlings below, darkening the sky with their numbers.

TOWNS

Contrary to popular belief, towns and cities offer a greater range of habitats for wildlife than much of the barren countryside. Man is a gregarious animal, living in such concentrations that his towns and cities are clearly visible at night from outer space, thanks to the large amounts of electric light used. He has created towns where birds sing through the night, foxes stroll unrecognised along the pavements in broad daylight, bats swirl around neon signs and badgers are wily enough to be hand fed – a land of super-lichens, super-rats, adaptable animals and opportunist plants.

Any town or city is made up of several separate habitats, all of which are eagerly colonized by plants and animals. It is quite likely that there are many creatures born into this urban situation that have never experienced the green countryside. Their behaviour, food and choice of breeding site may, therefore, be completely different from their contemporaries living in the wayside and woodland.

The inner city is closest to being the concrete jungle of which people speak. Jungle is perhaps an appropriate word since there is plenty of wildlife to be found there – hardly surprising when you think of the many situations directly analogous to the countryside. The vertical sides of high rise buildings, for example, are just like natural cliff and quarry faces to breeding birds, complete with little ledges for their nests. Moreover, they can expect little interference from man for most of the time, apart from the odd window cleaner! Golf courses provide an ideal habitat for orchids and other interesting wild flowers, whilst other large, open spaces make good resting places for flocks of migrating birds. Suburban gardens offer a mosaic of native and introduced trees, shrubs and other wild flowers, as well as homes for a host of insects, birds and mammals. Similarly, the open spaces of parks, large gardens, ponds, rivers and boating pools provide not only a recreation area for people but further habitats for wildlife. Each different area harbours a whole new collection of wildlife on your doorstep.

TOWN BIRDS

Many different groups of birds have exploited the urban environment, moving in from woods, heathlands and coasts in search of food and breeding sites. Many familiar birds of the town garden are truly woodland species. Blackbirds, titmice (blue, great, coal and long-tailed), woodpeckers, treecreepers and nuthatches are all birds which normally feed on the fruits of the forest – seeds and berries, caterpillars and beetles – but, thanks to their great powers of adaptation, they have entered the urban environment with great success. They can still be found in the woods of the countryside but the appetites of many have now been thoroughly whetted by the spoils of town life.

Several birds are common in towns and cities. The gregarious starling, swarming in its thousands to roost on some high-rise block, must be high on the list. In fact, so common have these birds become in the cities of Western Europe that we have invented deterrents aimed at reducing their numbers. They congregate in city centres, making an intolerable crescendo of noise at their roosting sites, and are messy feeders and nesters. They monopolise the bird table, rip the nut bags in their eagerness to get at the food inside, and gobble up any scraps with such speed that their rivals hardly get a look in! Compare this behaviour with that of their country counterparts and you would be forgiven for thinking that the two were entirely different birds.

In the countryside, starlings tend to alight in small groups and fan out to comb the ground in a most efficient manner. They also follow sheep, and even sit on their backs whilst looking for parasites – a trick also learnt by the magpie – or merely probe the ground for earthworms and leatherjackets. Unlike solitary species such as blackbirds and robins, starlings seem to be most successful at finding food by working in small family groups or larger flocks. They also derive safety from predators by feeding together.

Apart from their noise and mess, starlings have caused other quite unusual problems. Flocks

TOWN BIRDS	
Blue tit	Grey wagtail
Bullfinch	Kingfisher
Chaffinch	House sparrow
Chiffchaff	Long tailed tit
Cole tit	Redshank
Collared dove	Redstart
Cuckoo	Robin
Dunnock	Starling
Greenfinch	Tawny owl
Great tit	Wren

occasionally alight on the hands of Big Ben in London putting out its time, and at their nocturnal roosts the birds have been known to snap branches from stout trees with their accumulated weight! Efforts to remove them from their urban stations have included taping and playing back their alarm calls, and putting a slippery paste on window sills and ledges to provide a less secure footing. Neither method has achieved great success in the long term, however.

For all their faults, starlings are lively antagonists and great mimics too. Their song is very variable as they impersonate other birds as well as mimicking noises heard about the garden – including human whistles and trim-phones! They have particular roost sites which may be several kilometres from their feeding grounds, and groups of them can often be seen heading homeward to the same place at the same time each evening. In suburban and country areas the roost site may be on a single electricity pylon, all the birds neatly crammed on to the cables and metal-work, each with its own individual space. Tens of thousands come together in these winter roosts which may also be in woods. During the summer, food is more abundant and the birds disperse into smaller groups.

Black redstarts occur in towns too. They are a Mediterranean species which migrates northwards into Western Europe each spring and

Left: *Starlings are opportunist scavengers found in almost all urban habitats. They have two broods each year and lay up to seven pale blue eggs in a nest of dry grass lined with feathers, often built on the roofs of houses or factories. Unlike the irridescent blue, green and purple plumage of the adults, the young are dull brown.*

Below: *Cotoneaster is frequently planted in gardens and parks for its wonderful display of colourful fruits during autumn. These shrubs are doubly useful for wildlife since the little flowers are eagerly sought by honey-bees for their nectar and the red fruits provide a rich supply of food for blackbirds and other birds later in year.*

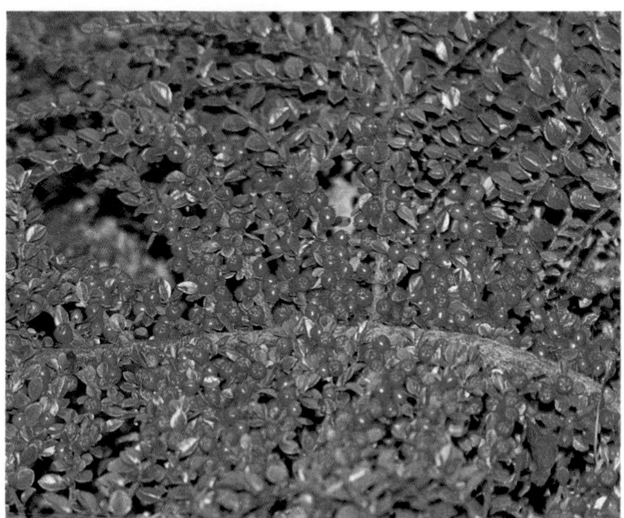

returns in the autumn. In their native area they are often found round houses and farms, nesting in sheds and outhouses as does the robin in Britain. Ever since man colonized the forest, the black redstart has exploited the nooks and crannies of his buildings. As if to emphasize this association, the black redstart is often called the house redstart in rural areas of Germany.

In towns and cities, these birds may be found on office buildings, and around factories, power stations and railway yards. Here there are plenty of nesting sites for them in odd corners and on sheltered ledges where they can make the nests of

WHY WILDLIFE IN TOWNS?

Why does wildlife thrive in towns and cities? This is the question often asked by those who see birds, insects, mammals and wild flowers burgeoning forth in urban areas. Ultimately, it depends on whether they have the 'genetic capability' to live in a totally new environment. It is a phenomenon called adaptation.

Not all wildlife can adapt to a new environment

The open spaces of town parks and gardens often attract scavenging birds like these black-headed gulls and feral pigeons. These and other birds such as sparrows, rooks, crows and herring gulls soon learn where free food hand-outs are provided by bird-loving people.

overnight. Only some species have the capability to cope with a new habitat. Others do not; they cannot exploit new situations. They survive only in their original habitat and become extinct when this changes. Those that do survive thrive, because there is less competition, more food and more shelter.

There are distinct advantages in living in the town or city if you can adapt. The temperature can be much warmer – up to 5°C higher than in the country – and the effects of this can be verified easily by simple observation from a train window during spring. The leaves will appear on some trees in the city whilst those in the country are still in tight bud. During winter, too, the urban environment is less cold. Some parks and squares ringed by buildings escape the rigours of severe frost altogether, so increasing the chances of survival for those creatures overwintering in these milder man-made areas.

Food is always in greater abundance in the town and city. The concentration of gardens in suburbia offers a refuge for many forms of wildlife. Man has planted millions of trees, shrubs and wild flowers in his gardens and parks, each a potential source of food for insects and birds. Millions of dustbins and rubbish tips are there to be raided, not to mention the wastelands – forgotten factory floors and open roofs burgeoning with wildlife.

Apart from the new environments created by man, forgotten corners of original habitats often exist right into town centres. Deep ravines and valleys formed by streams and rivers escape the developer's hand. Here wildlife thrives as it has always done, secretly out of the gaze of urban man.

THE HOUSE SPARROW

Of all the European sparrows the house sparrow is the one most associated with man. The hedge, tree, Spanish and desert sparrows follow in decreasing order. The house sparrow is a small and very drab bird but this belies its enormous success. It is believed to have spread across Europe from Africa during the Stone Age and can now be found wherever there are human settlements.

One of the most successful birds throughout Western Europe is the house sparrow. Its strong association with man makes it one of the most familiar birds of the urban environment, being found in gardens, housing estates, factories and car parks – anywhere, in fact, where it can find waste food and suitable nesting sites.

After courtship, the adult birds will raise up to eight chicks, taking it in turn to incubate the speckled eggs. The young soon leave the nest and the parents prepare to raise further broods – they may have as many as four in one year.

The young, like the female of the species, are rather drab, light brown birds, lacking the black bib, pale cheeks and white wing bar of the male.

The house sparrow is one bird that almost everyone is familiar with, thanks largely to its perpetual chirping, its precociousness and its sheer bravado in front of man. Today, it is found in towns, villages, farmyards and city centres from North Africa well into the Arctic circle and throughout East and West Europe. Its small size enables it to exploit holes in roofs, thatching, factories and warehouses. These birds have been known to make their nests entirely out of glass-fibre where this is readily available – quite a change from the hay used when nesting on cliffs or in thickets – and will readily use bits of string, paper and plastic to consolidate it. The sparrows' diet is just as flexible, including grain, seeds and insects as well as large quantities of bread and scraps. They are gregarious birds eating and feeding together, and are now an inseparable part of nearly all city centres.

grass and feathers in which they rear their young. Isolated power stations in coastal areas often attract migrating redstarts, thanks to their numerous ledges and walls which make ideal landing and breeding sites. They can often be seen fluttering from ledge to ledge, jigging up and down in pursuit of small insects – very common and likeable birds to have around!

Apart from common birds like the starling and sparrow, the urban environment sometimes plays host to a score of more unusual visitors. Some cities, of course, are luckier than others, according to their proximity to other habitats or their geographical position – in relation to flight paths and so on. In Aberdeen (Scotland) waxwings can be seen throughout the city gardens feeding on the fleshy fruits of cotoneaster and whitebeam, especially in those years when there is a big migration of the birds into the country during autumn. In that city, too, rare ivory and glaucous gulls have been seen scavenging on city lawns.

At some ski resorts the resourceful snow bunting has turned to a seasonal life of cashing in on the throw-away food inside crisp packets and

BIRDS OF THE TOWN AND CITY

It is surprising how many unusual birds may be found in town and city centres, yet most people remain unaware of them. Coastal towns are often graced with visits from seabirds or migrant species which come to rest after their long journey. The higher temperature of such environments can also result in birds usually associated with warmer climes surviving in the wild. Escaped budgerigars and parakeets are good examples.

Black-headed gull. *This gull is now a very common sight in towns and cities well inland. Its resourcefulness at scavenging for food – whether in tips, parks or docks – has led to it establishing some large urban populations.*

Oystercatcher. *A coastal wader, this bird has taken to roosting on rooftops in seaside towns, and searching for invertebrates on nearby playing fields and other open spaces. They also breed on relay structures set up for the North Sea oil industry.*

Above: *Greater flamingoes feed on plankton in the shallow salt lagoons just behind the coastal town of Palavas in Southern France. They are part of a colony which breeds in the Camargue Regional Park but can be studied best when at their feeding grounds close to the habitations of man. They appear oblivious to the humdrum traffic of major roads nearby and can also be seen in water adjacent to Montpellier airport.*

Black redstart. *These summer visitors to north-west Europe are common throughout the southern part of the continent. Their nest sites are frequently on high-rise blocks or power stations where there are plenty of undisturbed, cliff-like ledges.*

Waxwing. *These are winter visitors to parks and gardens where they feed on berries, particularly rowans.*

Ring-necked parakeet. *Following accidental as well as deliberate introduction, these tropical parrots have set up large colonies in some urban parks.*

sandwich wrappings, a role usually undertaken by the house sparrow. Being a bird of rocky and mountainous areas has not stopped it from making great adaptations to the incursions of man into its territory. Similarly, the city of Montpellier in Southern France is graced with frequent visits by greater flamingoes to the stretch of water adjacent to its airport. These extraordinary birds get their bright colours from the algae filtered from the shallow lagoon water in which they feed.

TOWN HAWKS

Uniquely amongst the daytime birds of prey, the kestrel has been successful at exploiting the urban environment. Larger raptors such as the buzzard, kite and harrier do not now include towns and cities within their territories in Western Europe. The larger hawks are much more sensitive to the presence of man and do not adjust easily to living alongside him. In less populated times, however, the kite did scavenge over London, feeding on discarded food left in the streets, as indeed vultures do today over the outskirts of eastern cities.

Kestrels, or windhovers as they used to be known, have found towns and cities just as good for food and breeding sites as their native countryside. They breed successfully in many city centres, in Birmingham, London and Paris for example. They have bred on Notre Dame in Paris and have been seen flying over St Paul's in London – a great contrast to the sight of small groups of them hunting in the boggy wastes of the National Park of the Auverge in France. In adopting city life, their diet has inevitably undergone a subtle change. In towns, food is more abundant than in the country, and it is different. Hovering and flying over the neat flower beds, squares and private courtyards in

Above: *Like many other birds, kestrels choose to make their nests on the precipitous ledges of buildings, often affording good views to office workers unused to such country delights. Buildings have largely replaced the cliffs, quarry faces and tree cavities which provide kestrels with their more traditional nest sites, although specially made nest boxes can also induce them to nest in the urban environment.*

Left: *The buzzard is a common hawk found throughout Europe. It has a strong association with man and his activities and is such a regular sight perching on the autoroute posts in France that special signs have been put up to inform motorists of its presence. At harvest time, these birds wait for mice, rabbits and hares to flee from the stubble. They rear 2-4 young in a bulky nest in a tree or on a cliff ledge. Occasionally a juvenile may stray off course during migration and be befriended by man, like the one shown here. They are not normally seen in towns.*

the city, kestrels are mostly on the lookout for small birds. They have been seen to swoop down and catch sparrows from garden lawns and to take other birds from bird tables. It has also been suggested that some town kestrels are becoming scavengers since they have been seen taking kitchen scraps. In the wild, however, a high proportion of their food would include small mammals such as mice, shrews and voles.

Apart from kestrels, sparrowhawks have also made a tentative venture into the city. These striking birds are now becoming more common in the countryside too, having been killed by the thousands during the 1940-60s due to the constant use of organic pesticides. Many have been found nesting in city parks, or seen taking food from bird tables in suburban gardens. Sparrowhawks are skilled fliers, feeding mainly on small birds, though the high incidence of feral pigeons in city centres has probably led to them taking these larger birds as well.

Strangely enough, it is the pigeon population that we have to thank for the introduction of yet another hawk into at least two European cities. Plans have been made to introduce several pairs of peregrine falcons to Paris in a grand scheme to curb the rising number of feral pigeons. Although previous attempts at West Berlin airport proved successful, it is estimated that there are about 15,000 pairs of pigeons in Paris with each making about eight nests each year and laying an average of two eggs each time. Thus the task of the peregrines to rid Paris of pigeons would seem to be an almost impossible one!

TOWN OWLS

Only a handful of the fifteen species of European owl have exploited the urban environment. The barn, tawny, little, scops and long-eared owls may all be found living in towns and villages, attracted there by the large number of rodents, small birds and invertebrates. Most of the town owls retain their nocturnal habits and hunt at night, although some will come out in the afternoon or early evening.

Originally a hunter of copses and well-wooded countryside, the tawny owl has now extended its territory to include parks, gardens and even refuse tips and wasteland. It hunts for its prey of

Above: *Tawny owls have taken up residence in many city parks. They are perhaps the commonest of all our owls but are rarely seen during the day, except when food is extremely scarce. They occasionally fly silently over city parks on the look-out for small mammals and will also take frogs, slugs, fish, worms and insects. This one has gone blind after hitting a car on one of its nocturnal flights.*

insects, frogs and small birds mainly by night, but shortage of food during nesting time may force it out during daylight. Tawny owls live between five and nine years, exceptionally up to twenty-two years as one ringed bird did in West Germany. Urban individuals have been known to attack man.

Barn owls, to a lesser extent, have also adopted suburbia as an alternative hunting ground. They normally choose sheltered sites under masonry, in roofs and belfries of churches, and have been reported as breeding on Notre Dame in Paris.

In the country they quarter the meadows at dusk searching for mice, shrews and voles, seemingly oblivious to man, and quite content to chase quarry close to noisy tractors. Estimates of the barn owl population in the countryside suggest that they have declined in recent years. One of the reasons for this may be the lack of man-made structures – old barns and buildings for example – which still allow access to these birds once restored. Moreover, many of their original roosts in old farm outbuildings and ruins are being replaced altogether by modern silos and warehouses. An old tea chest firmly secured to a beam in a barn or outbuilding may serve as an attractive artificial nesting site for these birds.

For many years the little owl was the subject of much controversy. Having been introduced to Britain from Italy during the early part of the last century, it became widespread throughout Europe and earned the disfavour of a growing number of gamekeepers and farmers who believed it to be the main predator of their game chicks. It soon came to notice, however, that these charges were largely unfounded – analyses of the birds' pellets revealing large quantities of insect remains and little in the way of feathered prey.

The little owls found in suburban areas of cities are beneficial birds with a wide range of food. One analysis of 2,460 pellets from many

OWL PELLETS

Owls cannot digest the fur and bones of the small mammals they eat. Instead, they regurgitate these remains in the form of pellets, made up of a compacted mass of fur and bones collected from their last three or four meals. The pellets look a little like droppings and are usually discarded at the roost or familiar perch points. If you search around the base of large trees or beneath beams in old farm buildings, you will often find a collection of such pellets – a useful guide to an owl's regular roosting or perching sites.

You can also discover what small mammals are present in an area solely by analysing owl pellets. Each small mammal – whether mouse, shrew or vole – can be identified by its cranium (that part of the skull without the lower jaw) and the design of the teeth. The bones can be teased out of the fur in water, cleaned in potassium hydroxide and whitened in hydrogen peroxide.

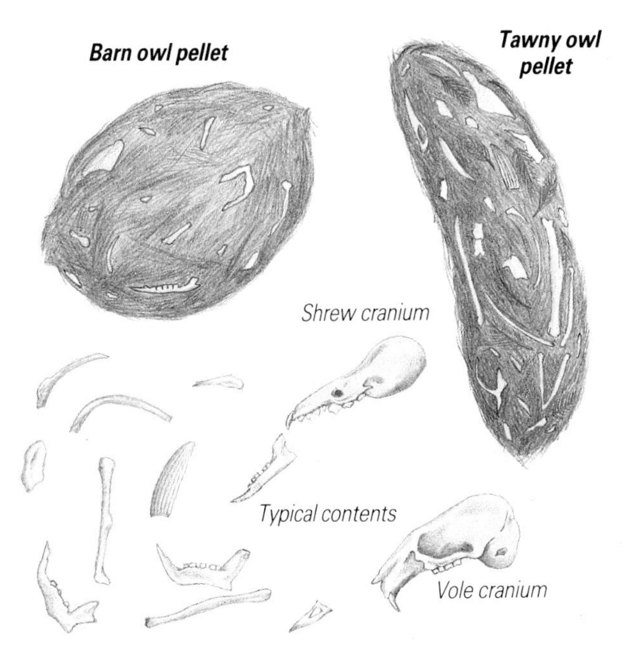

localities showed that about half of their food comprised of earwigs, cockchafers weevils, crane-flies (whose larvae are a great pest of the farmer) and ground beetles. There were few bird remains, including starling and sparrow, and the remainder was taken up with slugs, snails, woodlice, centipedes and millipedes. In other words, the little owl is an all-round scavenger whose diet remains much the same whether in town or in country. Unlike the other owls it may sometimes be seen perched on telegraph poles or wires in broad daylight.

TOWN PIGEONS AND DOVES

Tourists to city centres are frequently amused by the feral pigeons which crowd into squares and gather in regimented groups along the ledges of monuments and buildings. These feral pigeons are indirectly related to the wild rock dove which once inhabited coastal cliffs, but their immediate relatives in the city were domesticated dovecote birds which escaped during the middle ages and soon spread throughout Europe.

The feral pigeon is a very variable species in appearance, but can be recognised easily by the two black bands across its wings when at rest. Some specimens are light coloured with distinctive bands, whilst others are almost black so making the dark bands difficult to see.

Feral pigeons are a common sight in the busiest of city centres. They thrive in urban areas largely thanks to their good relationship with man who, despite risking a fine in some instances, continues to feed them with scrap food. Some plucky birds will venture inside buildings, pedestrian subways, railway stations and even underground tube stations. They solicit scraps at open-air cafes, sometimes taking food offered on the hand, the males strutting around cooing in their characteristic manner when courting their mates.

Studies have shown that during winter their food is almost entirely made up of bread, cake and bird seed supplied by man. In the spring and autumn many leave town to feed on seeds and stubble from the surrounding countryside. Another reason for their success in the urban environment is the vast quantity of undisturbed nesting sites on ledges of buildings in city squares, gutters, docks and parks.

Above: *Unknown through most of Europe earlier this century, the collared dove spread from the Balkans in the 1930s to Italy, Scandinavia, Germany and the British Isles from 1955 onwards. It thrives in suburban towns and around farms.*

Collared doves build a thin platform of twigs either in a tree or on a suitable roof which holds their two white eggs. The birds show a high degree of tolerance to man and will approach much closer than a woodpigeon. Pairs of birds spend much time together and will roost side-by-side on television aerials.

The collared dove is another very familiar town bird since it has exploited gardens and urban areas as well as farmland. Its spread through Europe over the last century has been quite dramatic. In forty years it has spread from South-eastern Europe through Scandinavia and the Netherlands to Britain. Following in the footsteps of its feral relatives, it will surely become a familiar sight in every city, town and village in the land. It is not an unfamiliar sight to see these doves in pairs for they seem to pair up for a long time, and scores of them can be seen resting on factory or farm rooftops. They make a characteristic purring noise and have a distinctive thin black collar which makes for easy recognition.

Left: Feral pigeons are a common sight in all European cities. Replete from their high street dinner, they can often be found resting only a few metres from busy streets and walkways. Gregarious birds, they feed, rest and sleep together for protection, highlighting the great variation in their colour forms, ranging from light to dark, patterned to almost plain. Some look quite obese, their crops being stuffed with food and therefore sticking out under their necks.

Right: A herring gull dozes off on the hot roof of a car in a town car park – a real sun-roof! Both adults and juveniles can often be seen benefitting from this somewhat unique form of underfloor central heating, resting on man-hole covers as well as vehicle roofs and bonnets. Note the orange spot on the bill of this breeding bird.

ROOFTOP ROOSTERS

There are a surprising number of coastal birds that have moved inland to adopt urban areas for breeding and winter quarters. Herring and black-headed gulls, for example, can be found nesting inland as well as in coastal towns. They feed on farmland, in rubbish dumps and parks, and are a familiar sight during autumn on playing fields and reservoirs far from the coast. They are traditionally associated with coastal areas and ports where their life is spent scavenging fish, taking food from other seabirds and pilfering young nestlings and eggs. However, their ability to eat food of almost any kind has led them to exploit man's environment far beyond the seaside resorts where they provide such entertainment during the holiday season.

Herring gulls often choose to nest among the stack-like chimneys around town centres where the rooftops are relatively undisturbed. They make a nuisance of themselves with their raucous calls, their untidy droppings and their dive-bombing of people who happen, often unwittingly, to trespass close to their young and nest sites. Gulls often defend their young vigorously with swooping fly-pasts – enough to frighten or terrorise anyone who does not know about these things. As one whose head (or hat) has been repeatedly hit by the talons and beaks of kamikaze-like gulls in the wilds of the Shetlands, it is no surprise to discover that their particular method of parental care still runs high in the urban environment.

Attempts to remove gulls from town rooftops have been relatively unsuccessful. Broadcasting their alarm calls simply has the effect of dispersing the birds to new nesting sites in the same area, as well as disturbing the peace of the town. However, it seems there is greater chance of physically removing a colony if it has not been established too long.

The herring gull is an opportunist with catholic tastes. Part of its success is that it is an omnivore, eating a great variety of food – from young birds, eggs, carrion, fish and crabs to vegetable garbage. At any quayside you can see it tucking into discarded fish and chips, eating crusts, thrown away sandwiches and other edibe refuse. It has learnt the trick of cracking pilfered birds' eggs by dropping them from a great height, and it delights in rubbish dumps where food is freely available. For this reason, it has thrived on the rubbish dumps of Marseilles on the eastern edge of the Camargue, to the detriment of other nesting seabirds. Colonies of terns, avocets and flamingoes have had their colonies pillaged for young and eggs, and weak migrant birds have been attacked and killed by being repeatedly whacked with the gull's tough bill. It is undoubtedly this strong survival instinct that has led to them becoming one of Europe's most successful species.

Three more coastal birds which have recently moved inland to some extent are the redshank, the oystercatcher and the kittiwake. Redshanks are common throughout Europe, wintering around

Above: *Greylag geese are the ancestors of the traditional farmyard goose. They are still found in the wild but have a much reduced distribution, more frequently seen in wildfowl parks, around lakes, reservoirs and gravel pits, often alongside Canada geese. They can also be found grazing in fields and town parks.*

the Mediterranean coasts although the British population is resident all year round. Normally birds of mudflats, marshes, moors and bogs, they have ventured up estuaries and rivers in search of food and in the last decade have been found feeding at least 10 km from tidal water. They have also taken to roosting on the roof tops of old town buildings. From here they go off to feed on the wealth of small invertebrates to be found in river shallows, at sewage farms and in flooded fields. It has been suggested that the lights from office buildings may assist the birds in feeding at night along urban river mudflats, although night-time feeding has also been seen in some wild birds.

Similarly, oystercatchers have always been traditional coastal birds. Now in at least one port, Aberbeen in Scotland, about thiry pairs of these handsome black and white waders nest regularly on the roofs of schools, hospitals and university buildings. The birds make rudimentary 'scrape' nests on the rooftops out of the gravel which covers the roofs for insulation. A mound is created into which a depression is formed to hold the eggs. The ensuing chicks risk being caught by city kestrels or, like herring gulls, falling off the tall buildings to their death. Despite this hazardous existence, the birds continue to nest in this alien environment, perhaps because the surrounding areas of carefully mown playing fields contain high concentrations of earthworms on which the oystercatchers have learnt to feed. In the wild, on coastal mudflats, they would normally search for marine worms with their sensitive orange bills, but their diet seems to have expanded from the molluscs and crustacea of their natural habitat to include inland worms and insects. They also visit other man-made habitats such as gravel pits and arable fields.

The pied wagtail is another species which has chosen rooftops as a safe retreat for its roosts. Originally a bird of freshwater margins and marshes, it too has adopted towns and cities for new feeding and breeding sites. Aggregations involving several thousand birds have been recorded in some urban areas. Groups of 1,000 to 1,500 have been seen on city rooftops and in busy tree-lined streets, as well as on factories in built-up areas.

Evidently the birds derive a lot of protection from these seemingly exposed areas – perhaps there is an advantage of warm up currents from extractor fans, a safe place to shelter from predators or a unique advantage conferred by

Above: *Normally birds of the open sea, kittiwakes have been found nesting in coastal towns. The ledges and window sills provided by buildings imitate the narrow cliff ledges where these birds usually make their nests. They can be distinguished from common gulls by their black wing tips and legs. Their name is derived from their screaming cry which sounds like 'kitt-ee-wake'.*

corrugated iron or asbestos. Whatever the reason it gives city naturalists an ideal opportunity to see more of this sprightly insect-eater, particularly since its nesting habits are changing to encompass ledges and cavities in buildings, sheds and walls.

URBAN MAMMALS

Just as the feral pigeon is arguably one of the most commonly seen city birds, so perhaps one of the increasingly common urban mammals also owes its existence to the fact that it was originally harboured as a domestic pet by man. Feral cats have been found in towns and cities throughout Europe for at least 1500 years. Today they can be found living in hospitals, gasworks, factories, docks and underground garages. They are communal by nature and groups of up to forty are sometimes recorded, though in most cities the average may be about seventeen. Their numbers vary from city to city – in one area in Birmingham a total of 349 were removed over a six-week period whilst the Colosseum in Rome is famous for its prodigious cat population.

The territories of feral cats vary according to the type of artificial habitat in which they live. In a country farmstead, the tom may patrol as much as 60-80 ha (150-200 acres) while several queens within this area may have overlapping

SWIFTS, SWALLOWS AND MARTINS

These birds, all summer visitors to Europe, are always being confused, yet there are striking differences between them. They all winter south of the Sahara in Africa and fly northwards in the spring. Some arrive in February and March if the weather is mild; the majority arrive in April and May.

Swifts have the largest wingspan and are the fastest fliers, often screeching low over built-up areas or soaring high over the countryside. They eat insects which get wafted involuntarily high into the air on wind currents. On a typical flight, a swift might collect about 1,500 small insects, packing them into a tight bolus or 'flyburger' which is then fed to the young.

Walkers in the country will not have failed to notice the swallow – a graceful flier with a long forked tail and bright chestnut throat. It flies along hedgerows and past walkers looking for insects disturbed from their refuge in the grass. It nests in many man-made places – on beams in stables, garages and farm buildings. The young have no trouble with their long tails whilst in the nest – they grow them only after they have left!

House martins have chosen the external walls and eaves of houses as their dominant nest sites. They prepare neat nests from mud collected at ponds and puddles, working it into a crescent-shaped bowl fixed to the soffits at the top of a wall. Originally, these birds nested on cliffs but ever since man started building houses, they have prospered on these artificial sites.

Sandy cliffs and quarries are the home of the sand martin. These birds are similar in size to the house martin but have a distinguishing brown breast band. They can be very numerous, particularly in old sand

Swallows always make their nests under cover, in this case between a rafter and the beam of a stable. Adults return regularly with insects for the young who will leave the nest after about 20 days. Swallows can often be found nesting in loose colonies and flying with martins, either in migration or when feeding.

and gravel quarries where banks have been left. Here they all nest together. It is easy to discover a colony of oval-shaped holes punctuating the top of a sand bank. In Southern Europe the holes can sometimes be confused with those of the majestic bee-eaters which also nest in colonies in sand banks.

Left: *A frequent sight in the town parks of southern England, the grey squirrel has replaced the native red throughout much of Britain. It is a cheeky import from North America which is quite at home in the presence of man – raiding the nut bag, taking food from the hand or collecting fallen sweet chestnuts from the lawns of city parks.*

Below: *Feral cats are always suspicious of humans entering their territories, and are much more wary than their domestic counterparts. This group exhibits a wide range of colour forms, from tabby to dark brown. Note the larger tom on the right of centre.*

territories, each about one tenth that of the tom. In the urban environment, where more food is available, the toms' territory may be as small as one twentieth of an acre, or a whole acre in dockland. Instead of their countryside diet of small mammals, small birds and insects, particularly grasshoppers, the feral cats scavenge on waste food like fish heads, left over fish and chips and potato peelings, as well as the occasional pest rodent which ventures into their territory. If the cats are befriended by a 'feeder' (someone who regularly feeds them), then more cats can be accomodated in a smaller area. Cats exhibit kin selection – that is related cats do things to help the common good of their group. They also have communal breeding sites where the kittens are reared and the cats sleep together, often touching each other for their group protection.

Many cat action groups have been set up in urban areas to counteract the problems from feral cats. Elimination of cats by trapping and killing helps to reduce their numbers temporarily, whilst some feral cats are returned neutered and ear-tipped which helps to restore social behaviour. Trapping of any animal does not always solve the problem, however, since some artful professional trappers often leave a few over to ensure continuing business the next year!

The grey squirrel is a familiar sight in towns and gardens. This agile mammal was introduced to Britain from North America in the late nineteenth century and has become well established here, although not elsewhere in Europe. It has made great adaptations to the urban environment, sometimes nesting in the roofs of houses, and is a frequent sight in gardens and parks. In some urban parks it may become quite tame, often feeding directly from the hand. In its natural habitat its diet consists of nuts, fungi, insects and even birds' eggs, but urban

Above: *The house mouse can be found living in houses, offices and factories, nibbling its way into cavity walls and beneath floorboards. Litters of five or six blind and naked young are born in nests of shredded paper, grass, leaves or sacking and within six or seven weeks they too can produce young of their own. In favourable circumstances one female may produce up to 10 litters a year. Mice droppings, which are about 5mm long, cylindrical and pointed at the ends, are often left on bare surfaces in the house.*

squirrels have been seen to accept bread and other edible refuse. They will store surplus food during the winter months by burying it in the ground and, although much of this is never recovered by the animal, the buried seeds and nuts often germinate to produce new tree seedlings the following year. Grey squirrels are active by day and do not hibernate – they may be seen feeding on fine days during winter, their brownish summer fur being replaced by a denser silvery-grey coat. They have few predators and can live up to ten years, although many die sooner as a result of starvation, collisions with cars or from pest control.

The red squirrel is the most widespread squirrel throughout most of Europe but it rarely ventures into towns, preferring heavily forested areas where it spends the majority of its time scampering amongst the treetops. Both species construct dreys during the spring and summer. These are usually made out of twigs and leaves around the fork of a tree or along a major limb. As well as overwintering quarters, they provide the squirrels with a safe place to rear their young. These bulky nests can be seen easily in bare trees during the winter.

Another familiar urban mammal is the house mouse. A pest in towns and cities for millenia, their nuisance value in the house is exacerbated by their producing more than fifty droppings a day, as well as a substantial amount of acrid urine. On farms they used to be responsible for making 10% of the grain in corn ricks unsuitable for milling, but modern methods of farming as well as more sophisticated poisons and traps have led to a reduction in their numbers on the land.

In cities, towns and villages, however, these mice thrive. They need at least 4gs of food each day, and existing side-by-side with man almost always ensures that it will be readily available. They will eat bars of stored soap, larder food, or even lengths of electric cable insulation which they gnaw at in order to wear down their front teeth which grow continually. Mice may be quick, but few are fast enough to escape electrocution when they bite too far through the cable!

House mice will live outdoors if the weather proves warm enough and food is amply available, but usually migrate to the shelter of houses and other buildings with the onset of autumn. Supposedly the commonest mammal apart from man, the house mouse is thought to have originated in Asia and began its association with man in the Stone Age. It is now found in almost all human settlements throughout the world. Unlike other species, the house mouse has a strong smell and a greasy brown-grey coat.

A more infamous guest in houses and factories is the brown rat. More common now than its history-making cousin, the plague-carrying black, the brown rat is, in fact, a very docile and clean creature – two reasons, perhaps, why they frequently make good pets in schools and colleges.

Both species are thought to have originated in Asia and have spread throughout the world with man's help. They are wily creatures and difficult to catch, being suspicious of any unfamiliar object. They colonize all manner of habitats and are good climbers, capable of running down slippery vertical surfaces with ease and even climbing up rope. The brown rat is also a good swimmer and can often be found around water, particularly near sewers. Both species inhabit many other places including overgrown gardens, outbuildings and hay ricks in the countryside, and warehouses, factories and other buildings in suburbia.

TOWN MAMMALS	
Badger	Grey squirrel
Bank vole	Hedgehog
Brown rat	House mouse
Common shrew	Noctule bat
Fox	Pipistrelle bat

Left: *A blind baby serotine bat, only a few days old, clings to the underside of its mother as she prepares for a night's foraging. Bats are mouse-like both in their size and coat, a fact which is reflected in both their French and German names –* chauve-souris *and* fledermaus *respectively, both meaning 'flying mice'. In Britain, it is an offence to disturb your own bat roost and to take photographs. This picture was taken, with authorisation, in the author's own home in which there is a healthy colony of serotine.*

Below: *Bats eat many moths on the wing, locating them with ultrasound and swooping in to scoop them up in their wing membranes ready for eating. Some moths can sense that bats are tracking them and will scramble the returning message or drop like stones to facilitate their escape. Large moths like the lime hawk shown here are taken readily by bats. Like the poplar and pine hawk moths, the lime hawk is named after its caterpillar's food plant.*

ATTIC BATS

Not the likeliest of mammals to find in the urban habitat, the bat thrives living alongside man – sharing his streets and houses, sleeping in the roofs of his houses, and hunting for insects above his gardens. There are over 25 species of bat in Europe, many of which will inhabit man's environment – using roof spaces, mines or old pit shafts as roosting places. Those most commonly found around houses include the pipistrelle, long-eared, whiskered, Brant's, Natterer's and Daubenton's bat. Roofs provide an ideal alternative to the bats' natural roosts in caves and hollow trees and they only need a gap of 20mm, opening to a space behind, in order to take up residence. Ill-fitting barge boards, ridge tiles, lead flashing, soffits and fascia boards are all likely to give access to attics and cavity walls. Warped weatherboarding, window frames and tile-hung walls also provide roosting sites for these beneficial creatures which feed on many insect pests.

It is unlikely that anyone with a resident bat will ever have the luck to study it at close quarters – they rarely appear during daylight and, once in flight, are experts at avoiding contact with anything but their next meal! Their amazing agility stems from their method of finding their way about – by sending out a stream of high-pitched squeaks and waiting for the echoes to bounce back from nearby objects. This system of echo-location enables them to distinguish between their moving insect prey and, say, a telegraph wire, and then to take the appropriate action. Unless one mistakenly flies through an

open window during cover of darkness and becomes trapped in the labyrinth of the house, the most you are likely to see of your local bat population is a dark shape flitting to and fro above the garden, local pond or, particularly in summer, the local sewage works.

Bats are protected in Britain where it is an offence to kill, injure or handle any species. They are inoffensive creatures and cause little trouble to the unwitting custodians of their roosts. They might sometimes be heard 'chattering' away overhead as they explore new areas of the roof, and baby bats may occasionally crawl in through an open window or arrive downstairs through a chink in the ceiling. So, to most, bats are a friendly addition to the house. They come out in the spring when insects are numerous – a pipistrelle may eat over 3,500 insects during one night, many of them garden pests – and continue their nocturnal flights until the autumn when they steal themselves away to hibernate in the cosy confines of our buildings and trees.

TOWN FOXES

People are frequently much endeared towards the fox. It is a handsome mammal which has the audacity to penetrate the suburban world of man, rummaging through our wastes and scavenging food from unexpected quarters. It seems equally at home along railway embankments, among the dustbins at backs of supermarkets, or even cheekily walking along suburban pavements in broad daylight. The fox is a beast of the woods and open country, but because of its stealth, slyness and wide-ranging diet it is now well associated with man. Its life in suburbia is less that of a hunter, more one of a scavenger and opportunist.

Urban foxes have received much attention from British scientists during the last decade, and what we now know from their work is true for urban foxes throughout much of Europe. These creatures are now a familiar sight basking on garage roofs, lurking in overgrown gardens, sitting with their cubs on railway embankments, or returning to their dens in wasteland and dockland.

Unlike some other urban creatures, foxes seem to avoid parkland and prefer the richer pickings of suburbia. They are quite content to wander among man's building and streets, and there is certainly no shortage of food waiting for them there. Like many other wild animals living in our towns and cities, their natural diet has changed to include man's waste food, as well as pigeons, blackbirds, sparrows, starlings, short-tailed voles and grey squirrels. Hardly surprising then, that the number of foxes in city areas is on the increase.

In 1983 a 'fox watch' organised in London recorded over 1,300 foxes in the capital and it may be that there are greater concentrations in cities than in much of the surrounding countryside. One reason for this could be that in such areas where food is abundant, foxes have a greatly reduced territory or home range. It can be as small as 40 ha whereas in the animal's natural habitat territories of over 200 ha are common. Indeed, in open country one fox may occupy as much as 1,000 ha due to the scarcity of food.

Below: *Foxes are familiar sights in towns and cities. They originally used the avenues of railways to penetrate deep into city centres. Today there are generations of foxes born into the world of dustbins, bird tables and litter bins that have never known the countryside of their ancestors. Urban foxes have to compete with cats and share their nocturnal prowlings. In some places they are brave enough to walk around undisturbed in daylight. Males have splendid chestnut coats, females are more drab.*

URBAN FROGS AND TOADS

It has always been good farming practice to drain the land. Straight dykes are cut through the fields and the water runs away quicker down these man-made ditches. But although this may make economic sense to the farmer, it sets many aquatic creatures at a disadvantage. Amphibians are one such group. The loss of these wet areas removes ideal breeding habitats, whilst dredging and herbiciding results in less areas of waterweed to protect developing tadpoles, leaving them susceptible to attack from predators such as herons and kingfishers. Many farmland and village ponds have also been filled in as part of the intensification of agricultural land, so with such a loss of habitat in the countryside it is not surprising that an increasing number of amphibians are being found in town and city gardens.

In the surrounding countryside, the main threat to amphibians is habitat loss. This means loss of ponds, lakes and ditches. Garden ponds, on the other hand, are not likely to disappear. In fact many more are likely to be made, thus increasing the habitats available for frogs and

Above: *The stripeless tree frog may be found around man-made wells and pumps as well as ponds and streams in Southern Europe. These amphibians exhibit extraordinarily good camouflage and can occasionally be found in vegetation several metres from the ground. In favourable localities 15-20 may be found in just one bush. The one shown here is a juvenile, measuring about 2cms in length. They grow to about 7cms.*

AMPHIBIAN METAMORPHOSIS

Metamorphosis or change in form is exhibited by frogs, toads and newts as well as insects.

Frogs start to breed early in the year, their spawn sometimes being seen as early as January. They will return to the same breeding place year after year, although pollution and drainage often forces a change. Without doubt, the commonest breeding site is now the garden pond where the frog's eggs will be laid in large groups. The black nucleus is protected by a transparent jelly through which the developing tadpole wriggles after about 14 days.

Common toads migrate to their breeding places in March or April, often tackling a variety of obstacles *en route*. Walls will be climbed and roads crossed, often giving rise to 'Toad crossing' signs as a warning to motorists. Unlike frog spawn, toad eggs are laid in long strings, about 2 to 3 metres in length. Both toads and frogs usually leave the water after spawning is over and live on land for much of the year.

Newts too spend summer and autumn living on the land. They move to their breeding ponds in spring and lay their eggs singly, wrapping each one in the leaf of a water plant.

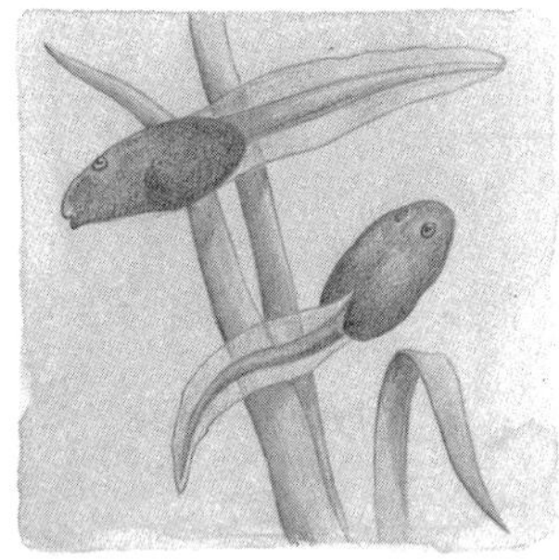

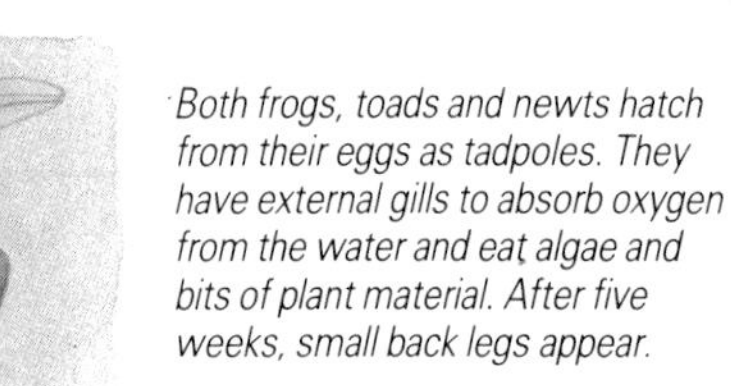

Both frogs, toads and newts hatch from their eggs as tadpoles. They have external gills to absorb oxygen from the water and eat algae and bits of plant material. After five weeks, small back legs appear.

After six weeks the tadpole will develop front legs as well. It dispenses with its external gills and begins to use its lungs. With frog and toad tadpoles, the tail is slowly absorbed into the body until, by about the time they are three months old, they are ready for life on land.

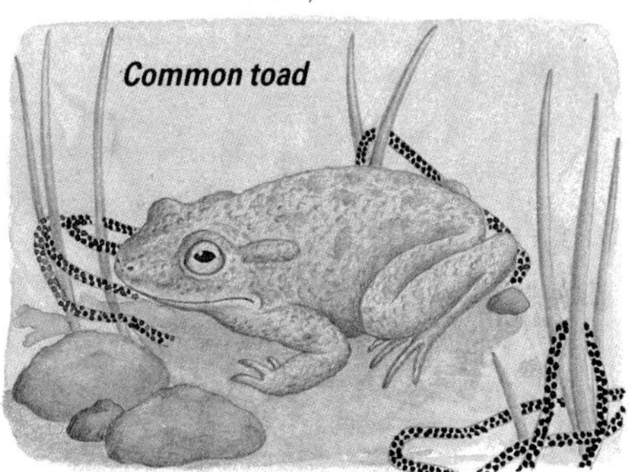

newts. Moreover, garden ponds do not frequently become polluted with nasty chemicals and detergents, and although suburban ponds and lakes may become extremely polluted when fed by streams which pass through industrial areas, amphibians often remain unaffected. This is either because they are deep in the roots of trees, below sand or in mud, or are away from the water altogether, inhabiting meadowland. I have often seen frogs spawn in park ponds alongside the detritis of modern man – throw-away beer cans, boxes and old bicycles. In fact, urban frogs seem to be doing very well. Reports are often heard of small ponds becoming overloaded with spawn, so much so that it is stacked up and protrudes from the surface of the water.

A greater problem for frogs and toads in suburbia is crossing the road! When the breeding season begins, these amphibians migrate across the land from their winter quarters to their breeding sites, and this often involves crossing roads and dual carriageways. They are not deterred from their original course however, and much consternation is expressed when they are squashed on the road in their hundreds. Intensive studies have shown that toads will travel up to 1.5 km (1 mile) to reach their breeding ponds, and frogs even further.

Despite their perilous existence in suburbia, frogs and toads continue to be exploited worldwide. These amphibians have been connected with pregnancy testing and diabetes experiments; in Latin America the skin of tree frogs is used to tip poison arrows; in China toads are dried; and in France frogs' legs are a delicacy. Hardly surprising then that although a toad may live to be over forty in captivity, in the wild only about 1 per cent live to be over eight years old.

Above: *Craneflies are a very important food supply for fish and birds. There are many different species which often emerge* en masse *in the autumn. They are attracted to lighted windows and can frequently be found indoors, frightening people with their long spider-like legs – hence their common name of daddy-long-legs. The larvae develop underground in damp turf and are commonly known as leatherjackets because of their tough grey skin. They are eagerly sought by rooks, crows and starlings in fields and are an agricultural pest. These two craneflies are mating which takes place soon after the female has hatched out. Sometimes there is a mass emergence of craneflies from mossy banks in woods and parks.*

INVERTEBRATES IN TOWN

Invertebrate is a general name which is used to describe a wide range of animal groups such as insects, spiders, crustaceans, slugs and snails, millipedes and centipedes. They all have one thing in common – they have no internal bone structure. Instead, they have an outer or exoskeleton.

Small creatures such as invertebrates have a distinct advantage in towns and cities. They are small enough to colonize cracks in walls and pavements, small crevices in buildings and spaces behind furniture and equipment. They may flourish unseen and build up large populations, especially those species which are active at night. You can find these creatures in all town and city gardens, no matter how little vegetation or bare earth is present. And in the house itself there are a hundred places where invertebrates – particularly spiders and insects – can find mini-habitats in which to live.

House spiders and harvestmen may take up positions on windows and ceilings, ready to catch any house flies which venture into their delicate webs. Green predatory lacewings are carnivores which can sometimes be found hibernating in large numbers in outhouses, sheds and attics. Cluster flies, which have a curious life history that starts off inside an earthworm, may also hibernate in attics during winter. In Southern Europe the praying mantis may venture accidentally into the house – a useful insect to have indoors because of the large number of flies it devours. Potter wasps, too, may make their fragile earthen pots outside on shutters. They fill each pot with a paralysed caterpillar and one of their eggs – the former as a ready supply of food for the latter's development.

Beetles in the house can be a nuisance, especially the woodworm, death watch and long-horned beetles whose larvae can live for several years on the nutrient-poor timbers. The devil's coach horse beetle may also be attracted indoors by the damp. You may see it running hurriedly across the floor, scaring people with its upturned body held high over its head. Earwigs, woodlice and silverfish are others that turn up in the house as unexpected guests. Woodlice arrive in from the garden looking for humid places to live, under door mats or damp skirting boards. Silverfish are wingless primitive insects which feed on organic matter – which often includes the paste used for wallpapering! They are so small that there are ample places for them to hide and, being nocturnal, they often have a free run of the house for their activities. Tiny book lice less than 2mm long, so named because they sometimes eat the bindings of books, may just as frequently be found on the bookshelf as in flour jars eating the nutrient-rich flour.

Turn on the lights when the sun goes down and a new hoard of insects will come to bombard the lighted windows. Craneflies, mosquitoes, gnats and various moths will flutter against the glass, not to mention large beetles such as maybugs or cockchafers which hit the glass with a startling thud and fall stunned to the windowsill.

Most of the invertebrates found in and around the house are native to Europe. Others, however, are exotic species introduced from the tropics. For example, at least two species of tropical ant – the Argentine and the Pharaoh's – thrive in the urban habitat. Both are about 2mm long, the latter being a pale orange colour, and are only ever found in buildings where there is constant heat – such as in central heating ducts, factories and bakeries.

Ants in the house are often a recurring nuisance. Their colonies are usually in places that are difficult to attack – in cavity walls, under walls and sills or outside, under paving stones. Mass emergence or swarming of winged ants in the house can cause alarm among its human residents, and attempts to eliminate colonies regularly result in failure. Ants only produce winged forms to fly to new sites. When they arrive they bite off their wings since they are of no further use.

Termites, too, have been a severe pest in at least four regions of Paris for the last 150 years. They have proved impossible to eradicate, despite evacuations of buildings and regular insecticide treatment. The threat to house timbers is great and the insects have been so successful that they have thwarted all attempts to eliminate them.

The house cricket, which is a native of Africa and the Middle East, is sometimes found in houses. In the North West of Europe it survives only in warm places such as bakeries and kitchens, but in the Mediterranean region it is

Right: *These polistes wasps are common in Southern Europe and make small combs on window shutters, under windowsills and in long grass.*

Below: *The mother of pearl moth, so named because of the irridescent quality of its wings, is often attracted to lighted windows. It is a common moth in the urban environment since its caterpillars feed on nettles. Each caterpillar rolls itself up in the leaves of its food plant when forming a chrysalis, holding them together with silk.*

TOWN INVERTEBRATES	
Centipede	Honey-bee
Clickbeetle	Hoverfly
Cockchafer (maybug)	Lacewing
Cranefly	Ladybirds
Devil's coach horse beetle	Millipede
Earwig	Mosquito
Flea beetle	Stonefly
Garden spider	Tiger beetle
Gnat	Woodlouse
Ground beetles	Woodworm

found regularly in old houses. As a nocturnal insect, it runs about at night on walls and sings with a high-pitched resonating sound. It has long antennae and spiny back legs.

Cockroaches are notorious pests in public houses and restaurants. They enjoy the sugary spills of beer and, if given the chance, will easily clog up beer delivery pipes. They breed fast in warm conditions and being nocturnal are not always seen until large numbers have built up. Even famous establishments get infestations of cockroaches occasionally – such as the popular case of German cockroaches in the Café Royal in London in 1983. The case against cockroaches as a vector of disease organisms to man has, however, never been proved. The two commonest species to be found are the common and the German cockroach. Both have colonized urban areas throughout Europe, surviving in warm buildings. They originally came from North Africa or Southern Asia but are now widespread thanks to man's global travels. They hide away in boxes and crates, infesting factories, warehouses and houses, and breed in cavities and heating ducts.

INSECT TYPES

With more than 30,000 different species in the insect class of Western Europe, it is convenient to group them into separate orders, each with special external characteristics. Those shown here are just a few of the commoner orders, examples of which often turn up in towns and gardens.

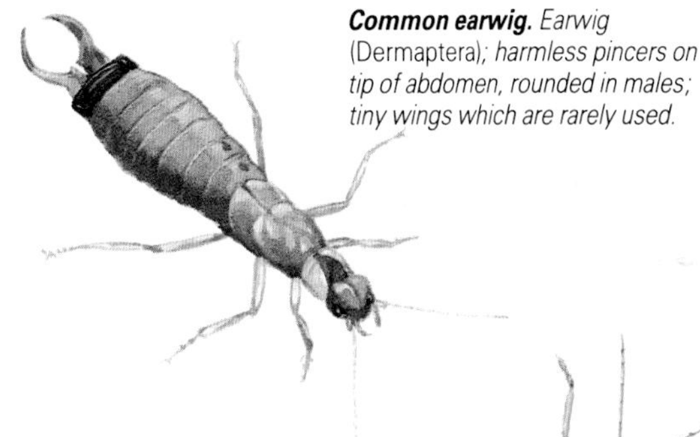

Common earwig. *Earwig* (Dermaptera); *harmless pincers on tip of abdomen, rounded in males; tiny wings which are rarely used.*

Two-spot ladybird. *Beetle* (Coleoptera); *fore wings modified to form hard protective wing case used for stabilising flight; sometimes fused together as in some ground beetles.*

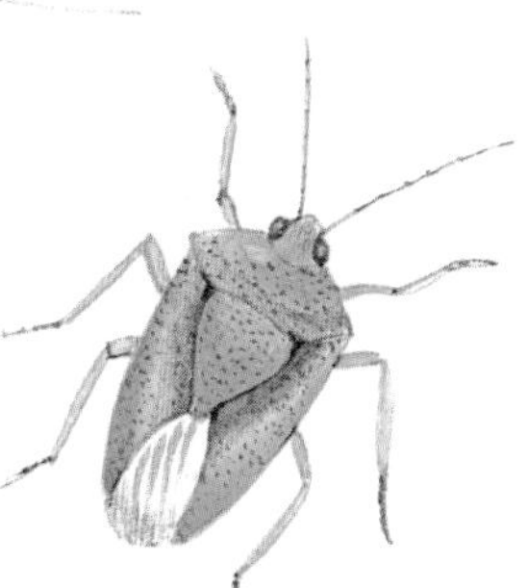

Black shield bug. *Shield bug* (Hemiptera); *shield-like fore wings which protrude at the tip; piercing mouthparts used for tapping plant sap.*

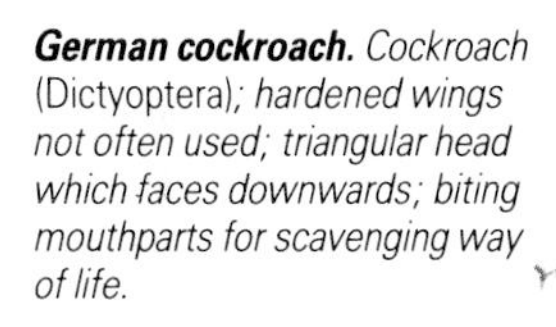

German cockroach. *Cockroach* (Dictyoptera); *hardened wings not often used; triangular head which faces downwards; biting mouthparts for scavenging way of life.*

Greenbottle. *True fly* (Diptera); *highly manoeuvrable pair of fore wings; hind wings reduced as gyroscopic organs for balancing flight.*

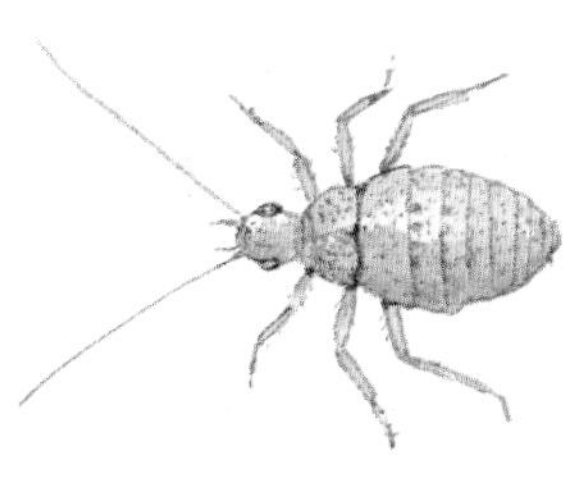

Book louse. *Louse* (Pscoptera); *tiny winged or wingless insects less than 7mm long; found in gardens, under bark or in houses eating food such as flour.*

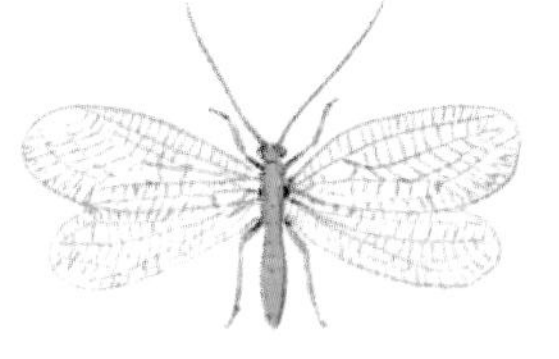

Lacewing. *Lacewing* (Neuroptea); *two pairs of membraneous wings conspicuously veined; large compound eyes and biting mouthparts befitting a predator.*

ARACHNIDS

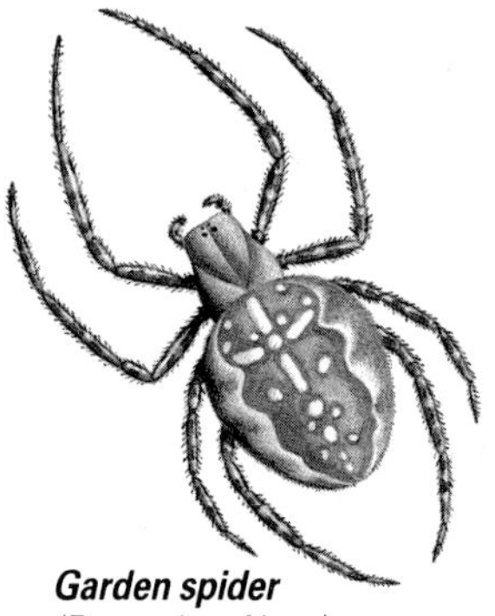

Garden spider
(Four pairs of legs)

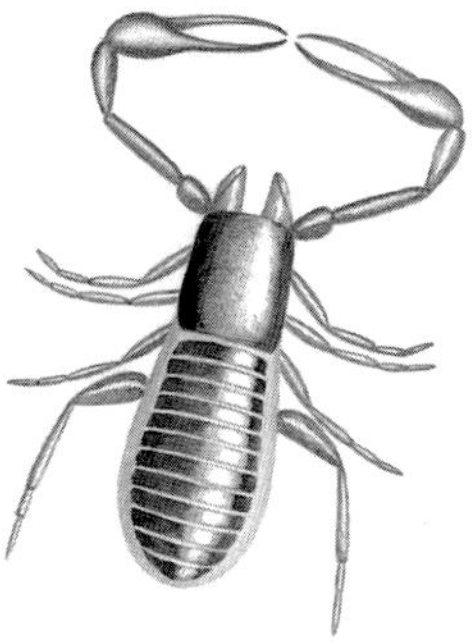

Pseudoscorpion
(3mm long; pincers)

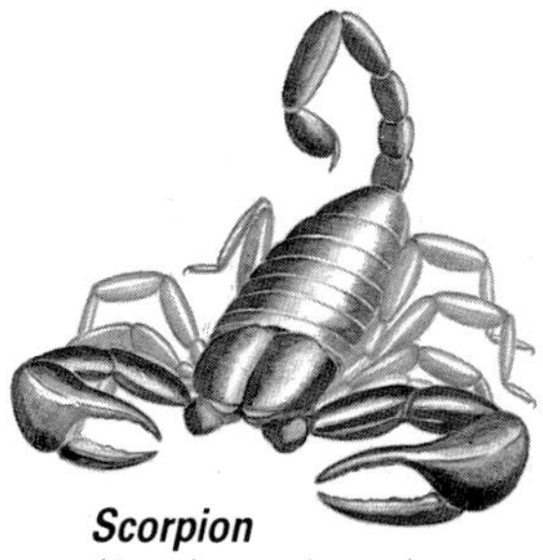

Scorpion
(4cm long; pincers)

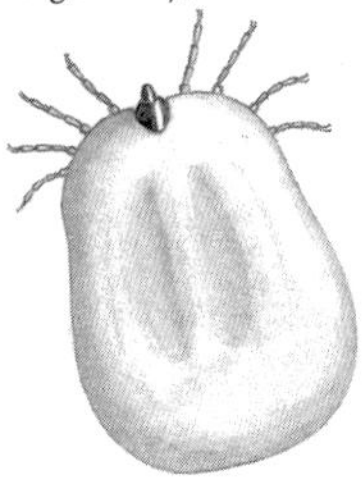

Tick
(5mm long; eight tiny legs; large body)

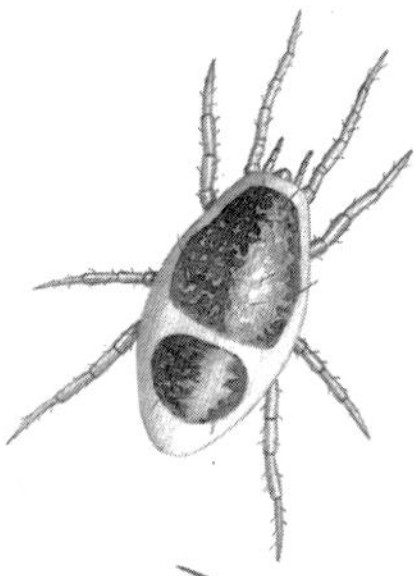

Mite
(Four pairs of legs)

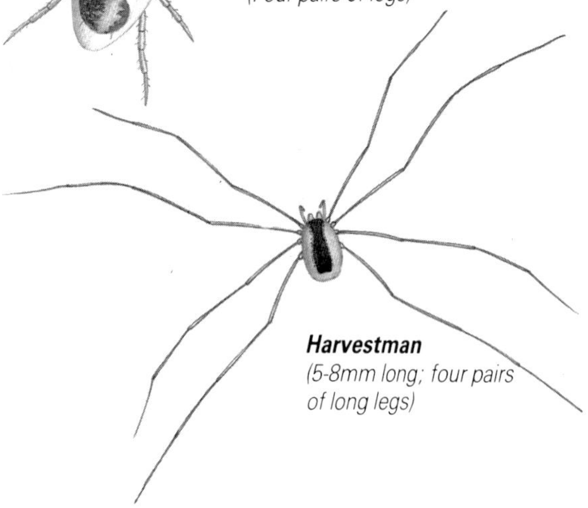

Harvestman
(5-8mm long; four pairs of long legs)

With over 700 species of arachnid in Northern Europe, the question of identification is problematic. Unknown species can sometimes be placed in their correct family by their shape (eg the crab spiders) or their type of web.

Characteristics of spiders include four pairs of legs and a body divided into two parts. Harvestmen, mites, ticks, scorpions and pseudoscorpions also belong to the class *Arachnida* since they too have eight legs.

SPIDERS

Spiders make a fascinating group to study. They colonize all types of habitat and it would not be unusual to find at least 20 species living in a small city garden. Some species are *only* found in houses – the familiar spiders that turn up with amazing regularity in the bath, sink and other odd corners – because of the dry conditions and abundance of small insect food. Other species have also exploited several man-made environments, such as mines and the concrete wastelands of motorway intersections. It is thought that at least one species, previously found only in wet meadows and sand dunes, has used its knack of being carried by the wind on fine silk threads to colonize these concrete wastes.

Spiders use a lot of ingenuity in capturing their prey. Some use webs, others rely on their stealth. Some, such as crab spiders, wait for their prey of grasshoppers, honey-bees and true flies, whilst concealed amongst flowers or vegetation. They have no webs.

Wolf spiders are hunting spiders which venture forth at night looking for prey. The lycosid tarantula of the Camargue in Southern France lies in wait for prey to come near its burrow before coming out and sinking its fangs into its victim. Jumping spiders, such as the familiar black and white ones found on the outside of buildings, stealthily approach their prey and jump on it at the last moment.

Then, of course, there are the spiders that use webs to catch their prey. The garden spider, for instance, belongs to a group which makes orb webs. As the name suggests, these webs are circular or semi-circular in shape and may festoon vegetation. They are used to catch flying insects such as hoverflies and true flies. In contrast there are spiders which make sheet or hammock webs which catch crawling or jumping insects such as froghoppers. These are common in hedgerows and, like all webs, become conspicuous in early morning when beads of dew stick to the delicate silk strands.

Other types of web and hunting techniques used by spiders in Europe include those which lurk in silk purses on the ground, waiting to kill insects which unwittingly walk over them; those that hang sticky silk threads from vegetation; and spitting spiders which project deadly strands of silk to cover their prey, effectively bringing them to the ground. There are also pirate spiders which surreptitiously kill other spider species by pretending to be a victim caught in their web. All spiders have a pair of tough fangs which pierce the hard shell of their prey ready for delivering the poisonous venom.

WASTELAND PLANTS

All cities and towns have derelict land – little pockets of undeveloped scrub, tiny oases of wild flowers set amongst towering buildings of concrete, brick and glass. The traditional bomb sites were a feature of many European cities during the first few decades following the Second World War. Most have disappeared now. Those derelict sites still in existence are usually where buildings have been cleared and leases are waiting to expire before redevelopment.

Look at any patch of derelict land and you will notice that the first colonizers are the plants. They find all sorts of unlikely places to live – sprouting out of bunged-up gutters, old walls and chimneys and around drain covers, prospering where plaster and pointing disintegrate, and seeding on the smallest patch of bare earth. On wasteland, plant colonization starts from the moment the soil is disturbed. Seeds carried on the wind fall into cracks and germinate; others which were previously too deep in the ground find themselves in the right place at the right time and begin the process of germination. Seeds blow in off the streets, get caught up in clothing and the treads of shoes or car tyres, and finally

Above: *Evening primroses are tall and conspicuous plants of wastelands and roadsides. Despite their name, they are no relation of the common hedgerow primrose, the only similarity being the colour of their petals. Forgotten gardens in cities, wasteland around docks, rubbish tips and factory walls are other places to look for this member of the willowherb family.*

Below: *The marsh woundwort is just one of a large family of plants thought by some ancients to be useful in staunching the flow of blood and binding up wounds. They thrive in the urban environment, preferring fairly wet areas as do their relatives the field and hedge woundworts which have similar pinkish flowers.*

URBAN BUTTERFLIES

Several butterflies have cashed in on the prolific amount of food available for them in the urban environment. Their caterpillars thrive on the opportunist wild plants that so swiftly colonize derelict land, forgotten gardens and overgrown cemeteries, the butterflies lapping up nectar from garden flowers and the beds of showy blooms in parks and other public places.

Amongst the aristocratic butterflies with rich colours and noble-sounding names, the small tortoiseshell is by far the commonest in towns and cities. Buddleia, which grows prolifically in waste areas, attracts the attention of this pretty insect, as well as the red admiral and peacock butterfly. All of these lay their eggs on stinging nettles – another plant which grows in abundance on derelict sites.

***Above:** The green-veined white is a common sight in towns during the spring and summer. It likes damp meadows and rough corners where the wild members of the cabbage family grow and on which it lays its eggs. Unlike many of its notorious relatives, the caterpillars of this butterfly do not attack the cabbages in the kitchen garden.*

Small and green-veined whites are also common in city centres feeding, as caterpillars, on wild members of the cabbage family. The large white can be seen in great numbers in some years, particularly when its caterpillars have been successful on kitchen garden cabbage!

The grass-feeding butterflies thrive on grassy banks and wasteland. Common and Essex skippers, small heaths, meadow browns and gatekeepers all lay their eggs on grasses. The striking colours of the common blue and small copper may also brighten a suburban garden if bird's foot trefoil and docks are present as appropriate food plants for their caterpillars.

A welcome sight in some city centres is the holly blue – a butterfly of gardens and open woodland. It is as common in some cities as it is in the countryside, breeding around well-established holly trees and ivy.

arrive on wasteland. When turn-ups on trousers were in fashion, quite a collection of grass seeds could be turned out at the end of a day spent walking through the country. From this, it is easy to understand how countryside plants can suddenly spring up in the centre of the city.

Colourful flowers on waste sites attract a lot of attention. There are several species which colonize bare areas in great numbers. One good example found throughout Europe is the London rocket, named after the Great Fire of London in 1666 when it quickly colonized the city. This plant can be easily recognised by its pale yellow flowers, overtopped by the long fruit pods which are often several times longer than their stalks. Its close relative, the hedge mustard, made a similar appearance on bomb sites after the Second World War. This plant also occurs along waysides in the countryside.

Another colourful wild flower of derelict sites is Oxford ragwort. It is named after Oxford in England since this is where it was thought to have escaped from cultivation at the end of the eighteenth century. It has been discovered, however, that it is actually a native of Italy, prospering amongst the volcanic debris on the lava slopes of Mount Etna. As with so many good colonizers, the rubble of stones and concrete in the urban environment simulates the plant's original territory – its Latin name of *squalidus* referring to its habitat rather than the plant itself!

OPPORTUNIST PLANTS

The wild plants which spring up on waste ground or disturbed land are opportunist ones. They produce plenty of seeds which are often distributed by the wind and may remain viable for decades, buried deep in the soil. If they fall in a suitable place, they germinate, grow, flower and seed very quickly. In urban areas they grow more quickly than in the country because of the warmer temperature.

Ivy-leaved toadflax. *A frequent inhabitant of old walls. It has ivy-shaped leaves.*

Mugwort. *This plant can be seen growing tall on roadside verges and waste areas.*

Purple toadflax. *Yellow is the normal colour associated with the toadflaxes but this variety is common in urban areas. Look for it in town and city centres.*

Wallflower. *As its name suggests, this plant is fond of walls wherever the mortar is friable enough to give it some purchase.*

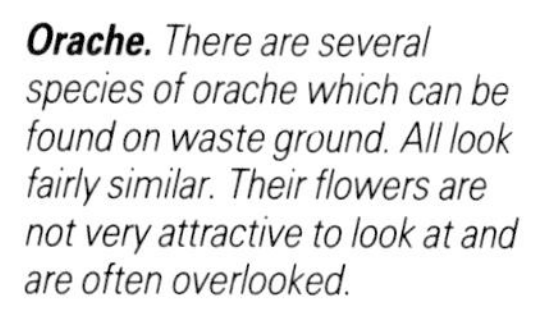

Orache. *There are several species of orache which can be found on waste ground. All look fairly similar. Their flowers are not very attractive to look at and are often overlooked.*

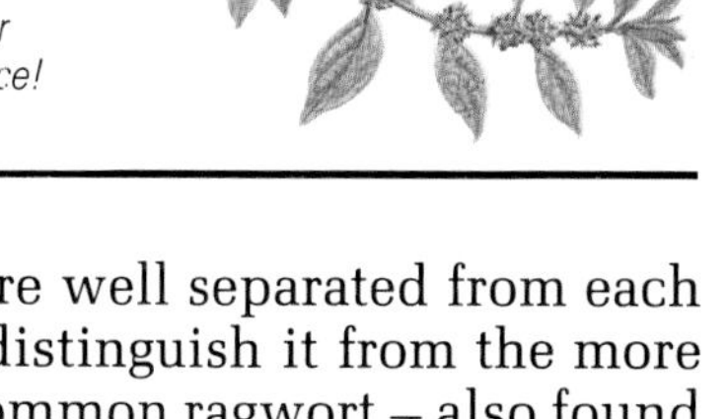

Pellitory-of-the-wall. *This plant is aptly named since it tenaciously clings to the base of old walls. It is used as a purgative and for cleaning bottles in rural France!*

Left: *Opportunist urban plants scrape an existence in the most unlikely places. Here wild antirrhinums or snapdragons sprout from a wall in the centre of a city, providing a welcome splash of colour amongst the concrete. They also occur in abundance along railway tracks in urban areas.*

Its yellow flowers are well separated from each other which helps distinguish it from the more compact heads of common ragwort – also found on wasteland.

Sticky groundsel is another alien, from North America, which frequents derelict sites. It is more of a floppy plant than its two ragwort relatives and has a sticky stem as the name suggests. A smaller but more familiar member of the ragwort group is groundsel, that pernicious weed of gardens which is such a good opportunist and ephemeral plant.

Ragworts all contain poisonous alkaloids, but these are tolerated by the yellow and black-

TOWN WILD FLOWERS	
Bugloss	Knotgrass
Burdocks	Mugwort
Buttercups	Nettles
Comfrey	Nightshades
Cow parsley	Ox-eye daisy
Dandelion	Ragworts
Daisy	Pimpernells
Fat hen	Snapdragon
Groundsels	Speedwells
Docks	Willowherbs

Above: *The Essex skipper, seen here feeding along with a 'bloodsucker' beetle from the nectar of an ox-eye daisy, lives in a variety of man-made habitats. It is very similar in appearance to the small skipper, the only difference being the darker undertip of the antennae in the Essex and the fact that it overwinters as an egg rather than a caterpillar. Though given a localised English name, it is widespread throughout Europe.*

banded caterpillars of the cinnabar moth which can strip tall plants down to their roots in a matter of days. The poisons are retained in the bodies of the larvae as a defence against birds. The adult moth has a different set of warning colours, red and black, and may be seen making ponderous flight over wasteland during the day. Burnet moths are frequently confused with cinnabar moths which breed on clovers, vetches and trefoils, sometimes on the same sites.

Old walls also prove a great attraction to many colourful plants. Wallflower, as its name suggests, finds sufficient moisture to survive and prosper on the barest of these man-made cliffs. Sometimes the mortar itself provides nourishment for plant growth but you can be sure that, however unlikely the prospect, every available niche will provide a home where some seed can germinate and grow. Snapdragon is another obvious plant which flourishes in clumps of red, pink and yellow throughout urban areas – particularly along railway lines.

Russian vine, or 'mile a minute', is one of many climbers that creep rampantly up old walls. Convolvulus is another which is difficult to remove once established. It produces few seeds but its extensive underground root system ensures vigorous vegetative spread. The funnel-shaped flowers are visited by numerous insects and often conceal large numbers of thrips – tiny insects also known as thunderbugs since they swarm on sultry days and land irritatingly on hands and arms.

In some places derelict sites may also harbour cultivated plants, survivors from the gardens that once existed there. Rockery plants in particular can often be seen tenuously clinging to walls and rubble, while rambling roses climb over old masonry in a blaze of pink and red.

It is the wild flowers on a patch of urban wasteland, however, that will give you some idea of how long that site has been abandoned. Just as on a motorway embankment, a succession of plants will take over from one another as the habitat becomes established. A new site may have Oxford ragwort, fat hen, orache and knotgrass which would gradually give way to goat's rue, golden rod, mugwort, rosebay willowherb, tansy and wormwood. (Tansy and golden rod are garden escapees which enjoy these conditions.) If the site is allowed to develop further, then a woodland habitat might become established; a sea of buddleia might give way to a canopy of sycamore, or a scrub of ash, sycamore, crab apple, goat willow, guelder rose, broom and Swedish whitebeam may develop. It is the natural order of events that plants succeed each other in the stabilization of the soil. Each has a temporary role in this struggle towards a stable or 'climax' community, which will develop differently on each kind of soil type.

Above: *The common twayblade orchid is one of the commonest to be found in towns, the others being the early purple and common spotted orchids. Twayblades do not have showy flowers like many of their relatives. Instead, their flowering spikes simply lengthen as the season progresses. They live in shady places.*

Below: *Recently disturbed soil provides a welcome space for many opportunist plant seeds. Dormant seeds are brought closer to the surface to begin germination. Others are blown or carried to the site by birds and mammals. Soon a piece of freshly-disturbed soil sprouts forth with poppies, corn marigold, mallow, wall wheat and many others.*

Other clumps of vegetation which are common on wasteland include knapweeds, teasels and thistles. Nettle clumps are always a sign of man's activity and signify nitrogen-rich soils. Melilots and mallows make attractive clumps of colour which can be seen from a distance, and tall spikes of mullein and evening primrose are very distinctive. The ground might be studded with hawkweeds, dandelions, daisies, plantains, fumitory, convolvulus, scarlet pimpernel, and wild strawberry.

INDUSTRIAL WASTELANDS

Unlikely as it may seem, industrial wastelands support a great variety of plants, insects, birds and mammals. They offer a wide range of different habitats such as old dumping grounds, quarries and mines into which flora and fauna have quickly become integrated. Urban sprawl has led to many estates being built on such sites, their gardens backing on to these areas of rough common land, deserted spoil heaps and man-made lakes.

You might not think that such sites could harbour a wealth of wildlife but, in many cases, they provide refuge for a huge number of species driven out from the surrounding countryside. In fact, some nature reserves have been created around interesting industrial sites. The type of waste – whether fuel ash from power stations, smelting slag, coal slag, tailings from industry or metalliferous waste from tin, lead and zinc mines – determines the type of soil. This in turn encourages certain types of plant to grow. Plants, of course, encourage insects which encourage birds, and so on in a natural cycle of events. A rich mix can be found right on your doorstep.

One of the best places to study wildlife in these refugia is on calcareous sites where lime-rich spoils have been dumped. The ground may be studded with orchids which thrive on this type of soil – over 100 flowering spikes per square metre – including species like the common twayblade, common spotted and early purple orchids. These are the three commonest orchids in Northern Europe and can be found in both town and countryside. Other less common orchids found on these industrial sites include the northern fen and frog orchid. In wetter areas the early and northern marsh orchid have been recorded.

At least one orchid and one bird owe their existence to the quarries and industrial sites of our towns and cities – the dark red helleborine, which is an orchid (not to be confused with hellebore), and the chough – a handsome black bird of sea cliffs and rocky headlands. If these last refuges are disturbed, their relic populations are likely to be severely endangered.

Abandoned machinery, chimney stacks, flues, shafts and ledges in quarries also provide nesting sites for many other birds – including ravens, kestrels, buzzards and barn owls – as well as several species of bat.

Many slag heaps offer a unique chance to study fossils since these relics of the past are often in a readily accessible position. The ancient coal measures – once a living wet forest – are sometimes very rich in fossilized plants and animals. Recent research into such sites has turned up fossilized fish, amphibians and plants from 300 million years ago. But the search for such remains has to be done before the slag heaps are landscaped or redeveloped, so liason with developers is necessary beforehand.

REFUSE TIPS

Many urban areas, and some countryside districts too, are graced with municipal refuse tips – containing the domestic rubbish from countless homes, shops and industries. Such rubbish

A GUIDE TO SOME WASTELAND BUTTERFLIES

Large aristocratic butterflies, whites, blues, browns and skippers are all at home on the waste ground of towns and cities. Whites will be attracted to wild members of the cabbage family found growing there, browns and skippers will come to lay their eggs on the grasses, and aristocratic butterflies abound where nettle patches provide their caterpillars with an adequate food supply. Small populations of common blues and small coppers may also arise where docks and trefoils are present.

Apart from prime egg-laying sites, the butterflies are also on the look-out for rich nectar sources, and many of the plants which flourish on wasteland are good providers. Buddleia, the butterfly bush, grows prolifically in urban areas, setting seed in all sorts of places. It often forms dense thickets on waste ground and its heady blooms become covered with visiting butterflies of all descriptions during the summer.

TOWN BUTTERFLIES	
Brimstone	Painted lady
Comma	Red admiral
Cabbage white	Small tortoiseshell
Green-veined white	Small white
Orange tip	Speckled wood

Small tortoiseshell. *One of the commonest butterflies in towns and gardens. Its caterpillars feed on nettles and the adult insects sometimes hibernate indoors.*

Meadow brown. *Very common in the long grass which grows on commons and in the rough areas of parks and golf courses. Both sexes have 'false eyes' towards the tips of their fore wings to scare birds.*

Large skipper. *An impressive butterfly which may be seen basking in the sun with wings swept back. It is always ready to mount an exploratory sortie against a fellow butterfly or other insect which intrudes into its air space.*

Above: *The shoo-fly plant, or apple of Peru, is a South American species which can turn up on rubbish dumps, sprouting from imported bird seed. It is a member of the potato family and its presence is said to deter flies.*

Left: *Black-headed gulls scavenge for rich pickings on an urban rubbish dump. They are often accompanied by black-backed and herring gulls, jackdaws and crows. They have also been seen to exhibit remarkable hearing, for when the rubbish vans arrive and make their monotonous piping sound on reversing, the sound acts as a signal for the gulls, hundreds suddenly arriving from nowhere.*

Large white. *Backyard cabbages are always under threat from the caterpillars of the notorious large white. Prevention is always better than cure so remove the batches of yellow eggs as they appear.*

Red admiral. *Migrations of these aristocratic butterflies move north-westwards through Europe during the summer. Its bright colours once earned it the name of the alderman, resembling as they do the livery colours of noblemen.*

Holly blue. *This butterfly is sometimes very common in parks and gardens wherever there are extensive ivy banks. The female has distinctive black and white striped antennae and legs.*

dumps are exciting places to explore because of the large number of wild flowers, reptiles, insects, birds and small mammals that they attract, but unfortunately they are an endangered habitat. Many are grassed over or tidied up for development by aesthetic-conscious councils and the ensuing loss to naturalists is great. Further irritation is caused by the fact that old tips and quarries soon develop a unique flora and fauna of their own, often supporting species found nowhere else locally. When councils wish to fill in or dump further rubbish, the naturalists naturally wish to retain their rarities! In some cases sites are designated as places of special scientific interest, often because of their closely knit populations of orchids and butterflies, and are left alone.

Plant colonizers of rubbish dumps vary from the common to the unusual. Cannabis, for example, is now a very occasional colonizer of rubbish tips and waste areas but how does it get there? The answer may lie in the dirty bird tray bannished to the dustbin. After the pet budgerigar or canary has spilt half its seed over the dirt tray, it is carefully removed, the seeds later turning up on the local tip where they germinate. Other species which have been carried in this way include canary grass, sunflowers, Fuller's teasel and shoo-fly plants.

A very common plant of waste areas and refuse tips is the rosebay willowherb – often lending great splashes of pinky-mauve to recently razed ground. Its American name of fireweed is apt in this respect because, though given for its fiery display of colour in late summer and autumn, the plant certainly encroaches upon the cinders and charcoal left by fires. One of the reasons for its success lies in the fact that it is a prolific seed producer, each seed wafting away in the wind on delicate feathery plumes. It also produces underground shoots which allow it to form large clumps, sometimes to the exclusion of other plants.

Refuse tips also attract seabirds in quantity. The pickings are so good that birds find it worthwhile to commute from the estuaries where they normally live to sites well inland – a round trip that sometimes covers as much as 113 kms (70 miles). The black-headed gull is usually the most abundant seabird at refuse sites, no matter what the time of year. It is also the commonest gull of parks and gardens where it scavenges for food put out for smaller birds. At the turn of the century it was a rare bird on the edge of extinction, then it changed its dietary habits and became a scavenger. Today, it ranks as one of the most successful birds in the urban environment! But beware of trying to identify it. During the winter

Left: *The common blue is really common, as its name suggests, on urban wastelands, and in the rough areas of parks, commons, roadsides and railway embankments where clovers and bird's foot trefoil – the food plants of its caterpillars – can be found. Here a pair can be seen mating. Note the darker female below and the diagnostic arrangement of black spots and orange lunules on her underwings. Only the male has the bright blue wings that are so distinctive in the summer.*

Below: *The brown rat frequently lives alongside the house mouse on refuse tips. Males can reach 30cms in length, their tails adding a further 20cms. Females are smaller.*

it does not have a black head – the feathers (which are in fact chocolate brown) are lost during its winter moult to be replaced by white ones. Then one day in spring you will notice that these birds have all got their black heads back again. Black-headed gulls as well as arctic terns have also ventured up the Seine to the centre of Paris scavenging for food.

Large refuse tips may also have resident groups of crows, jackdaws and rooks over the winter. They forage for waste food and congregate on trees adjacent to the tip at dusk, often making the trees look black. Rooks nest in tall trees, often near buildings and in city centres. These rookeries are usually permanent homes, the birds returning to the same nests each year, and may contain anything from several hundred to a few thousand nests.

Apart from the larger species, many smaller birds of the open countryside have found the habitat of the refuse dump to their liking. Common species such as sparrows, starlings and pigeons are joined by corn buntings, reed buntings, skylarks, tree pipits, wheatears and whinchats who have all discovered that breeding close to such sites can work distinctly to their advantage. Some birds have made adaptations to include edible refuse in their diets but those that remain purely insectivorous are attracted by the large quantity of insects found in such places. There

are often numerous butterflies and caterpillars, many species of fly and beetle, not to mention grasshoppers and other denizens of the tip. House crickets can also be very numerous, with populations running to 5,000-6,000 individuals.

The two commonest mammals to be found living in refuse tips are, predictably perhaps, the brown rat and the house mouse. Many other mammals, both small and large, may visit the tip out of curiosity or hunger but few choose to make their homes there. The mice population decreases in winter when the mice either take refuge in buildings or remain as a small overwintering group, surviving deep in the litter and debris.

CHURCHYARDS AND CEMETERIES

Churchyards and cemeteries are great refuges for wildlife in urban areas and sometimes represent ancient habitats. They are certainly rich in wild flowers, trees, lichens, insects, birds and mammals.

Churchyards are much older places than cemeteries and therefore more interesting wildlife might be expected to be found there. The great quest in studying the botany of a churchyard is to see if the wild flowers growing there indicate a really ancient site. The site could have been enclosed from adjacent meadowland or permanent pasture several hundred years ago. If so, the chances are that it will not have been ploughed or herbicided, so a rich relic site may have developed.

Naturalists frown at very prim and proper churchyards. They also frown at overgrown jungles. The happy medium is to create a semi-wild pasture which is still rich in wildlife, though a small area of overgrown land is useful for the earths of foxes and setts of badgers. The use of weedkiller, reseeding and regular mowing to make neat lawns is no good for wildlife. If some sort of control over grass and scrub is necessary then it is best to graze or cut after June when the seeds of wild flowers have set. The removal of tombstones to the church side or wall not only disrupts the variety of lichens likely to be found, but is an irreversible step for local historians and archaeologists. Many churchyards and cemeteries provide schools with local areas to study natural history and, indeed, many schools and conservation groups carry out active work in these areas. In fact, there are some churchyards and cemeteries which are now specially designated as nature reserves because of their wildlife interest.

The most obvious feature of an overgrown

Right: *Churchyards are a paradise for botanists, often carpeted with flowers like the lesser celandine or pilewort seen here. They have mostly escaped the hand of man and are rich in trees, lichens, wild flowers, insects, birds and mammals.*

Many abandoned churchyards and cemeteries have been made into urban nature reserves since they have been left to run wild and now support their own rich collection of wildlife. They are havens for birds and mammals which make regular sorties into the surrounding urban gardens and parks. The peace and quiet of the thickets which spring up are just what they need. Leaving nature to its own devices is one of the best ways of encouraging wildlife.

WILDLIFE IN CEMETERIES

There is more wildlife in many towns and cities than in much of the surrounding countryside. This is because they provide a great many refuges – churchyards, parks, golf courses and, most important of all, gardens.

Overgrown cemeteries are often lush with dense vegetation; avenues of trees once planted as formal walkways now tower loftily overhead, providing nesting sites for pigeons, shelter for hundreds of insects and welcome support for climbing plants such as ivy. The temperature in towns and cities is a few degrees warmer than in the country, which extends the growth period of many species. Hardly surprising then that such fine specimens of flora and fauna may be found living alongside man.

1 Sycamore
2 Lombardy poplar
3 Elder
4 Ivy
5 Xanthoria lichen
6 Small heath butterfly
7 Bramble
8 Wren
9 Seven-spot ladybird
10 Leaf mine
11 Hogweed
12 Common shrew
13 Speckled wood butterfly
14 Convolvulus
15 Red campion
16 Woodpigeon
17 Harebell
18 Foxglove
19 Lime hawk moth caterpillar
20 Common lime

cemetery or churchyard is the invasive nature of the trees. Mixed in with the formal shrubs planted by graves and tombs as a mark of respect, are the native and non-native opportunist trees and shrubs which have colonized the area by themselves. Sycamore and silver birch are often found in great profusion on old sites. Ivy seems to tackle all the trees in some areas and envelopes the trunks and branches in a rich cladding of leaves and stems. In the suburban environment much cooing of pigeons can be heard from the dizzy heights of tall trees. The birds find the entanglement of ivy an ideal place to nest. Avenues of limes and horse chestnut, planted as formal walkways through cemeteries last century, have matured over the years to support a vital population of insects, and laurels, privets, snowberry, holly and rhododendrons have gone wild, pushing up to great heights or straggling through the undergrowth.

There are a few types of tree which turn up with great regularity in churchyards. Yew is one. It is said that the wood was used in the making of longbows, although some writers believe it was only the Normandy yews that were suitable, growing to a much greater height than the English ones. Pollarded at 2 metres from the ground, these would not have provided sufficiently long wood to make the shafts. Yews may also have been planted as evergreen windbreaks. There are some churches in Britain where yews are planted on the south-west side of the main entrance thus acting as a break against the dominant south-westerly winds. In days when more hats were worn perhaps this was a practical thing to do – after all, you didn't want your fine hat to blow away just as you were entering the church. The most convincing argument for yews in church-

***Left:** The Monterey cypress is a North American species introduced to Western Europe in the last century. Now well-established in many churchyards, the trees provide ample food for the caterpillars of a moth called the Blair's shoulder knot.*

yards, however, is that they were planted as a symbol of respect for the dead. They are the most long-lived trees in Europe, some estimated as being over 2,500 years old. Some become hollow – there is at least one in existence in Britain which used to have a tea house inside – while others develop a huge girth and seem to split up to give the appearance of several yews growing together. The yew's resinous bark is not an inviting habitat for lichens but the space between the branches offers nesting sites for blackbirds and thrushes.

Another tree seen in churchyards is the Scots pine. Clumps of these have traditionally been planted, not only in churchyards but in the countryside, to designate high points and the extent of estate boundaries. In Britain at least, where secrecy over religious faith and the chance of losing your head ran high, this tree was also planted by an incumbent with Jacobite sympathies. The strawberry tree was also used as a code for the upholders of some particular heresy. Widespread in heath and maquis, this tree has hairy red fruits rather like strawberries.

Mulberry trees are also found in religious places, particularly in the precincts of cathedrals. They are mostly the black rather than the white species which is most suitable for the caterpillars of the silkmoth. A lot of the mulberries in Britain originate from cuttings of those planted during James I's campaign to get the British silk industry started. This was induced by the immigration of several thousand French Huguenot weavers into southern England early in the seventeenth century. Today many of these mulberries – all of which turned out to be the wrong species for the silkworms – may be seen in the precincts and grounds of monasteries, churches, abbeys, religious colleges and schools all over the country.

Churchyards are also wonderful places to look for wild flowers in the spring. Carpets of lesser celandine may be mixed with wood anemones, primroses, cowslips, bluebells, snowdrops and daffodils. Many species were probably originally planted at gravesides by mourners but the plants soon colonized the churchyard and offer a colourful spectacle when mixed with other wild flowers.

Soon, the spring flush of colour gives way to the flowers of early summer such as cow parsley, creeping buttercup and hogweed. If you are

CONFUSING UMBELLIFERS

White flowers which look like umbrellas in shape (hence their family name of *Umbelliferae*) cause a lot of confusion in identification. If trying to name such a plant, it is important to note the habitat in which it is growing and the time of flowering. Height can also be a good guide. The umbellifers attract a lot of insects (which come to them for nectar and pollen), including true flies, hoverflies and beetles.

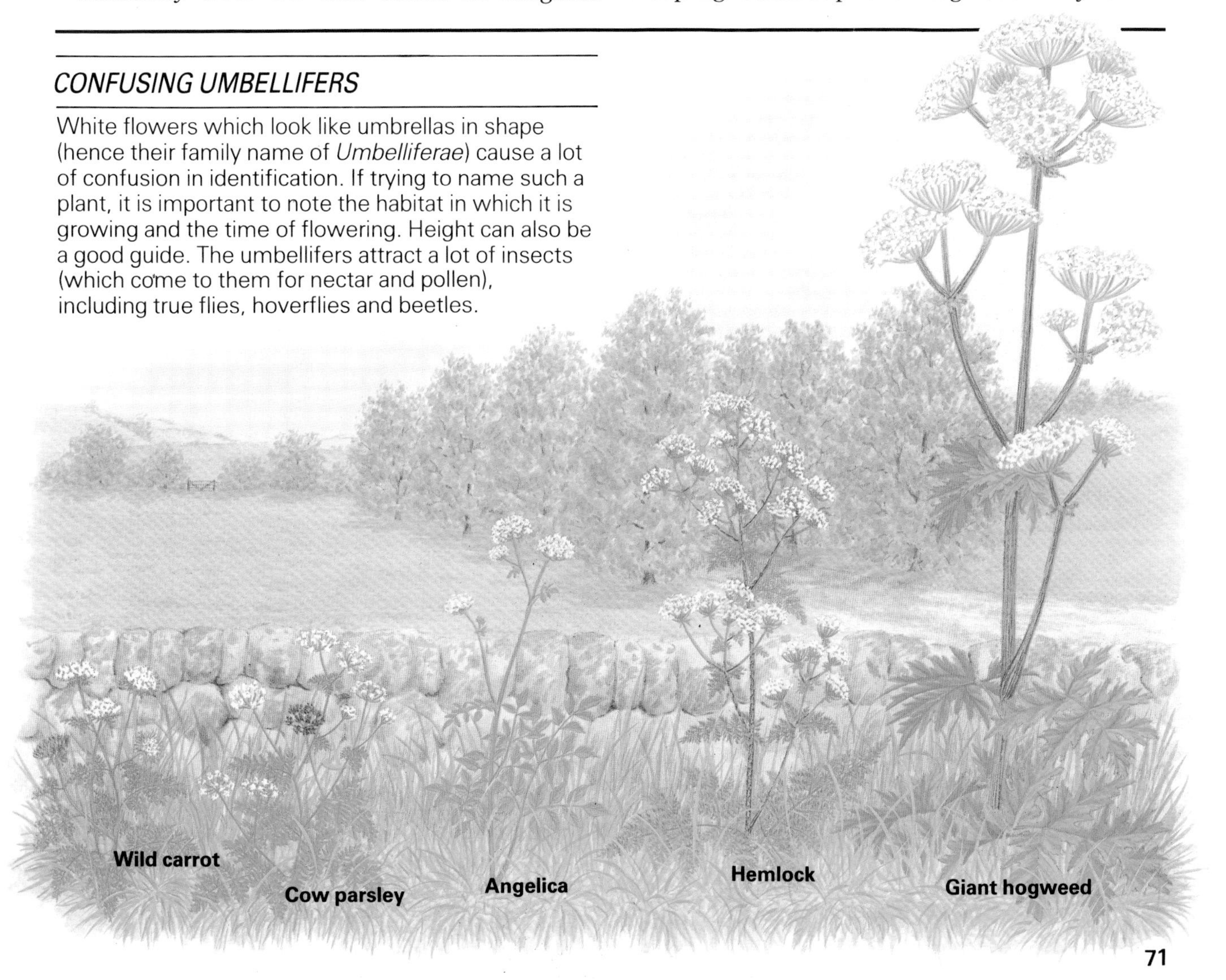

WHAT IS A LICHEN?

Lichens are plants formed by a symbiotic relationship between two partners: an alga and a fungus. The algal partner contains chlorophyll thereby allowing photosynthesis to occur. In turn, the sugars made during photosynthesis give energy to the lichen.

There are over 2,000 lichen species in Europe, found in almost all habitats from the very cold to the very hot. Many are highly sensitive to pollutants. Their slow growth is stunted by pollution and they may die altogether if there is too much sulphur dioxide in the air. They suffer from the dilute sulphuric acid (a part of acid rain) which falls on them when sulphur dioxide and rainwater combine. This pollutant is produced during burning, so high concentrations tend to occur in city centres, thus creating 'inner lichen deserts' where, until a few years ago, lichens were completely absent. In the suburbs only resistant lichens were able to survive so these areas were called 'outer lichen deserts'.

The presence or absence of certain lichens is used by lichenologists to give a measure of relative pollution in and around cities. In recent years there has been a recolonization of lichens in some city centres, partly due to a concerted effort to reduce the amount of pollution through 'clean air' acts. Anyone travelling into a city by train can still notice, however, that there is a gradual decrease in the number of lichens on rooftops lining the track as they near the city centre.

Lichens also absorb pollutants in the most unlikely places. In Lapland, cow's milk was found to be high in radioactivity – this was traced to the cows eating lichens which had absorbed this aerial fallout. Since the nuclear test ban treaties this cycle has declined.

Lichens are excellent pollution monitors – the more polluted an environment, the less lichen species will be found and vice versa. Some lichens sport these peculiar stalked cups which are part of their reproductive system.

Minute spores are violently ejected from the surface of the cup and are then distributed to new sites by the wind.

Above: *One of the most common lichens to be seen in Western Europe is the bright orangey-yellow* Xanthoria. *It occurs on rooftops, walls, farm buildings, monuments and gravestones and grows best where its substrate is rich in nitrogen provided by bird's droppings.*

lucky there may be bugle, yellow archangel, ground ivy and scarlet pimpernel, even burnet saxifrage and wood avens. Ox-eye daisy and bitter vetch are commonly seen in church pastures and knapweeds, thistles and nettles in rather more unkempt corners.

The diversity of vegetation attracts an equally wide range of insects and other animals. Skipper butterflies, meadow browns and small heath butterflies dance about on the flower heads and craneflies lurk in the damp depths of the long grass; frogs and toads may also be found there. Hoverflies patrol the flowers, and carnivorous lacewings and ladybirds lurk in the foliage. On a summer's day the sound of bumble bees working the vetches, clovers and trefoils is part and parcel of the average churchyard meadow.

The physical structure of many churches provides plenty of scope for nesting birds and insects. Honey-bees have established many permanent colonies in churches, under gables or in cavity walls, where they are completely safe from inquisitive beekeepers and from which their spring swarms regularly issue forth. Hibernating butterflies – small tortoiseshells and peacocks for example – find a constant temperature high up in the roof timbers or on the windows where they may stay for as long as nine months over the winter period. Pigeons and starlings use the gaps in the masonry on the outside for their nests, and barn owls use the open access to bell towers or attics for rearing their young – bringing in small mammals caught in the wild churchyard, surrounding countryside or urban waste. Moreover, colonies of bats are a common presence, hanging in the eaves and hawking above the churchyard on their nocturnal foraging trips.

Above: *Abandoned nests may be found in the undergrowth around churchyards. This one has probably been used as a feeding place by a wood mouse, hence the growth of plants from the uneaten seeds.*

The grounds of cemeteries and churches also provide nesting sites for meadow pipits and skylarks. Both birds fashion nests of old grass, roots, moss, and hairs on the ground, well concealed from prying eyes amongst the grasses.

Perhaps one of the most fascinating areas of study in cemeteries or graveyards is the lichens. These crusty coverings of colour can be found growing on gravestones, church masonry, walls, posts and the bark of some trees, and nowhere will you find such a diverse collection as in a churchyard. One of the most obvious is bright orange and is called *Xanthoria parietina*. Unfortunately, there are no common names for most of the lichens since they have never received much popular attention.

Over 80 different species of lichen have been found in some churchyards. Some like the tops of gravestones where they are regularly drenched in birds' droppings, enriching them with nitrogen. Others prefer the rougher surfaces of limestone, while few are found on polished marble surfaces. There are generally more lichens found on the sunny south side of a churchyard than on the north. Lichens do no harm so there is no necessity to remove them from gravestones. They enrich the natural heritage of an area and add colour and variety to a churchyard. Furthermore, they grow at an exceedingly slow rate – from half a millimetre to 5mm each year, depending on the species. To a certain extent they can be used to estimate the age of a tombstone or piece of masonry – by knowing what increments are put in each year, their size can serve as a basis for dating their origins. But beware, some lichens fall off and new growth occurs on the same spot. Similarly flakes of the substrate may split off. In Greenland there is a lichen which is claimed to be 4,500 years old!

Walls are also excellent places to hunt for different kinds of lichen. One, called *Lecanora muralis* (*muralis* meaning a wall), used to be found on rocks and boulders in mountainous terrain but is now exceedingly common throughout Southern Europe on man-made asbestos roof tiles, concrete, cement and walls. It forms a yellowish-brown crust. Another one can colonize concrete (*Lecanora dispersa*) and is accredited with being able to provide 85% surface cover in five years. It is found in city centres and is recognised by its white to grey colour. The other, called *Lecidea lucidis*, is a yellowish-green lichen found extensively on brickwork throughout the south east, particularly in damp shady places.

Some lichens have disappeared or become severely endangered due to changes in man's building materials. Two species (called *Lecanora farinearia* and *Lecidella pulveracea*) have become extinct due to the lack of old barn timbers and untreated wooden posts which were used widely last century. Most of today's posts are creosoted and offer little refuge for lichens. The *Cladonia* lichens which normally thrive on old thatching are very much reduced in numbers from the last century and the discontinued method of mud-capping walls has probably contributed to the decline of *Lecania nylanderii*.

Current threats to lichens occur where there are too many people; the trampling and scuffing of lichens at beauty spots, whether they are in upland or lowland sites, and from trampling and knocking at ski resorts.

TOWN MOTHS	
Angle shades	Garden tiger
Brimstone	Heart and dart
Cabbage moth	Silver Y
Codling	Winter
Elephant hawk	Yellow underwing

Parkland is one of the best places to discover wildlife. Many have been in existence for a considerable number of years, untouched by the rigours of agricultural development and urban sprawl. Thus, they tend to be rich in wild plants and insects, even ancient meadows. Many contain mature trees which supply sufficient decaying wood for fungi, tree-dwelling invertebrates and a multitude of nesting sites for birds and mammals. Unrestricted access to many public parks gives everyone the chance to study wildlife at close quarters.

PARKS

Parks and public gardens are important reservoirs for wildlife within towns and cities. They are distinct habitats by themselves. Some harbour plants and animals that you would only ever find in parks, as well as other forms of wildlife that use these green sanctuaries as resting and feeding areas between their forays into the concrete jungle.

Man has made parks – nothing quite like them occurs naturally. They are green islands in busy cities which offer a refuge for many wild things. In particular, they offer one habitat that is in short supply elsewhere in the urban environment – – grassland. Man has also made parks more attractive to wildlife by introducing trees. A mixture of native and foreign species, planted for their ornamental value, offer food, shelter and nesting areas for birds and mammals, breeding sites for numerous insects, and fruiting areas for colourful fungi. So whether the park is a formal affair, full of flower beds and neat lawns, or a rambling stretch of grassland, full of wild corners and scrub, it has great potential as a home for many diverse species.

Many parks in Europe have ancient origins. Some may have originally been enclosed from royal pastures or deer parks, and many of those in cities are famous in their own right – Hyde and Regents Park in London, the Bois de Boulogne in Paris, the Tiergarten in West Berlin. In general, the older the park, the more wildlife you can expect to find there. The same is true of the trees – the more old, decayed and rotting trees and logs there are, the more fungi, insects, birds and wild flowers you will find.

In even the smallest town park much activity can be recorded. Trees provide roosting sites for bats as well as many birds, and the ornamental flower beds and spectacular displays of bedding plants are a great attractant to insects. Butterflies, for example, visit flower beds for their day-to-day nectar intake. In fact, some municipal parks have actively encouraged butterflies by planting flowers specifically for their nectar content. Apart from feeding, the butterflies are also induced to breed in specially-left rough areas

Above: *The wood mouse, also known as the long-tailed field mouse, is found in many habitats, including gardens, parks, moorlands and even mountain-sides. It can be distinguished from the house mouse by its white underside, sandy brown coat and larger ears, and feeds on a varied diet of seeds, berries, shoots and buds, as well as snails and a variety of insects.*

Active only under cover of darkness, the wood mouse is a fast mover, running and bounding across open spaces. It is also an agile climber and frequently uses abandoned birds' nests as feeding places.

which harbour the nettles, grasses and other 'weeds' so vital to their caterpillars.

Bees, wasps, hoverflies, beetles and spiders are other small creatures which appear in abundance. Dependent on them are flocks of birds. Common species such as blackbirds, starlings, sparrows and pigeons are joined by the more unusual or summer visitors. Spotted flycatchers and hoopoes search for insects among the bushes, migrating waterfowl may stop at the park's ornamental lake and, at night, tawny and little owls swoop silently over the flower beds.

Parks also act as sanctuaries for large and small mammals. In the wilder areas, badger setts and foxes' earths may be well established out of the gaze of the ordinary citizen. Fallow deer – descendants from some Norman or mediaeval deer park – may scrounge sandwiches in the car park, and grey squirrels may pester visitors for tit-bits. Nocturnal hunters, like the weasel or stoat, may creep amongst the undergrowth looking for their prey of voles and mice, and rabbits reappear at dusk to resume their constant cropping of the grass.

If the park is graced with an ornamental lake, boating pool or meandering river then the mere presence of water greatly increases the chances of seeing something different – a dragonfly, a heron, the dazzling flight of a kingfisher or suburban dipper, or simply the collection of waterfowl that usually ornaments such places: mallards, moorhens, coots, even the occasional Canada goose.

PARKLAND TREES

Parks are one of the best places to study trees. Magnificent specimens tower loftily overhead, encompassing a whole world of wildlife within their branches. Unfortunately, the people who design parks rarely see the results of their labour in maturity. They create instead a joy for others; great avenues of trees, planted for their shape and colour, aesthetic landscapes where native trees cross branches with introduced species. Parks often contain trees from many continents – from Mediterranean tamarisks to North American redwoods and Australian eucalyptuses.

You can see how the trees were managed long ago by reading the evidence in their shapes. Many trees have large girths and look like huge stumps cut off at about 3 metres above the ground. This shows the practice of pollarding – the regular cutting of trees and shrubs to encourage new shoots. With pollarding, the cutting was done at a certain height above the ground to prevent grazing animals from eating the new shoots. The resulting poles were then cut regularly and used for basket-making (particularly willows), for fuel or for cricket bats in some species. Avenues of pollarded trees look quite

***Above:** The bright autumn colours of the rare service tree are occasionally seen in parks, its dark bark peeling off the trunk to leave light strips, thereby earning it the alternative name of the chequer tree. Its edible fruit was once commonly sold in markets.*

***Left:** The practice of heavily pruning trees is widely employed throughout Europe. Species like the London plane shown here lend themselves well to this apparently severe treatment which stimulates them to push out vigorous shoots the following spring. This provides a thick canopy of branches which confers plenty of shade to the street below during the hot summers of Southern Europe. Such regular pruning or shredding of trees often results in some intriguing shapes as exhibited by these town trees in La Salle, France.*

PARK TREES	
Ash	Monterey cypress
Elms	Mulberries
False acacia	Redwoods
Horse chestnuts	Scots pine
Judas	Silver birch
Limes	Swamp cypress
London plane	Sycamore
Magnolia	Tulip
Maidenhair	Walnut
Maple	Yew

attractive and are a common enough sight in many European towns and cities. Single specimens were once used as permanent markers or to designate boundaries. Wherever you find a pollarded tree, there is bound to be a considerable amount of wildlife associated with it. There are more nooks and crannies in its bark than in a normal tree, providing crevices and holes where nuthatches, woodpeckers or treecreepers can nest, as well as daytime resting places for night-time moths.

It may be that the parkland trees served another commercial purpose in years gone by – they may have been coppiced. This serves the same purpose as pollarding but involves the tree being cut off at ground level. An old coppiced tree today would not have just one single trunk; instead there would be several, all sprouting from the same 'stool'. Again, the principle of coppicing is that after a decade or so the new shoots are harvested to provide poles for fencing or pulpwood. One of the most familiar trees to be coppiced is sweet chestnut – or Spanish chestnut (which hints at its origins) – a species supposedly introduced to Northern Europe by the Romans.

Parkland trees such as oak and beech would normally have branches that come right down to the ground, were it not for grazing animals. Deer and cattle strip leaves and twigs to the maximum height they can reach. Thus, in parkland you can easily discover whether there are grazing animals about by looking for this neat cut-off line 2-3 metres from the ground. The effect is very striking from a distance. Should you come across parkland trees where there is no grazing to prevent the limbs touching the ground, there will be a bonus in the wildlife found there. Long grasses push up through the lower branches and make inpenetrable tussocks where small mammals, insects and fungi abound.

Many parklands are planted with trees which look showy – the magnificent flowers of magnolias, the curious blooms of tulip trees, or the wispy flowers and fruits of the smoke or wig tree. North American maples are very popular, their leaves turning gorgeous colours in the autumn.

WHICH SEX IS IT?

Have you ever wondered why a particular tree – like a yew or holly – never seems to produce any berries? If you have such a tree in your garden it is probably because it is a single-sexed or dioecious plant. If it is male, it will only produce pollen – no fruits.

Plants can roughly be divided into three different groups – those that have male and female parts in the same flower (hermaphrodite), those that have separate male and female flowers on the same plant (monoecious) and those that have male flowers and female flowers on separate plants (dioecious). There are several variables, however, including species with self-pollinating mechanisms and those which also have hermaphrodite forms.

Common single-sexed plants include red bryony, Canadian pondweed, asparagus, dog's mercury and winter heliotrope, as well as holly and yew.

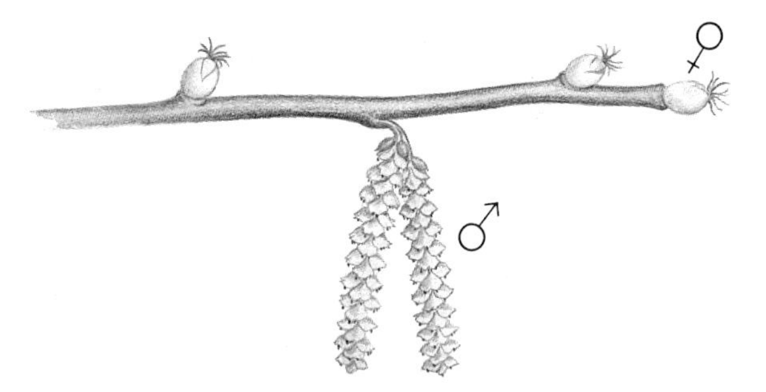

***Hazel** (monoecious). With this shrub, both male and female flowers are born on the same plant, often close together. The male catkins are obvious to the onlooker but the feathery red parts of the female are discreet and often overlooked. Where both sexes are represented on the same plant, it is a disadvantage if self-pollination takes place as it invariably weakens the species. For this reason, such plants have males becoming ripe before females or vice versa.*

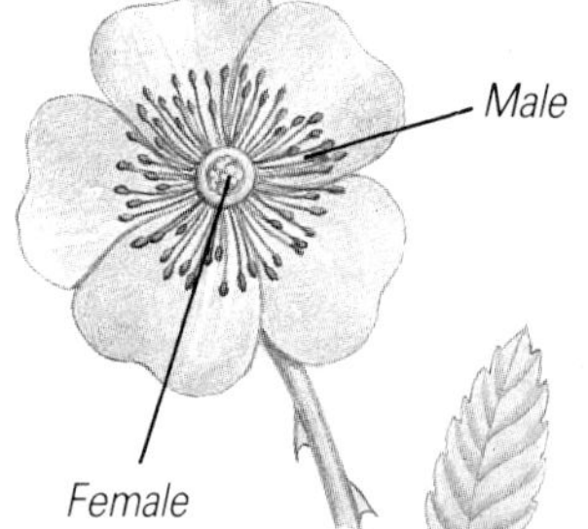

***Dog rose** (hermaphrodite). Each flower of this rose contains both male and female parts. The female parts are located in the centre – each made up of a stigma, style and ovary. The stamens of the male are arranged around the female parts.*

***Dog's mercury** (dioecious). In this plant ('dog' standing for common), male and female flowers are born on separate plants. The male plants may be recognised by their long, vertical tassels, and the females by their collections of little 'pom-poms'. It is named after the Greek god Mercury who is said to have discovered it.*

Californian redwoods are another common sight. With their spongy, flame-proof bark they push up to dominate the scene – always the tallest vegetation wherever they grow. Consequently they are the first to be caught by lightning. Look about your urban environment and see just how many redwoods have been decapitated – pruned by the weather. Afterwards they just sprout side branches and continue relentlessly upwards.

Monkey puzzle trees are another familiar sight in parks and public gardens, yet they are comparatively recent arrivals in Western Europe. They were brought over from South America in 1791 – as a result of the collector Archibald Menzies pocketing some unfamiliar nuts that were passed round at a meal! Another curious tree met in parks is the maidenhair tree or *Ginko biloba* – a relic tree known from fossils. It is a single-sexed species and pairs are often planted together.

Modern parks are often rich in smaller trees too. The forest species such as oak, elm and lime are too big for many of the newer urban parks since they can grow to 20 metres in height and cause much damage to adjacent buildings. This is the great problem facing today's urban developers – a legacy of huge forest trees planted in our towns and cities by bygone town planners. Replacements in modern parks tend to be small species, with showy flowers in the spring, showy

Above and left: *Parkland trees are often planted for their showy bark, flowers and fruits. The London plane (left) has attractive patchy bark and has adapted well to the polluted urban environment.*

Sweet or Spanish chestnut (above) is a common parkland species with magnificently showy flowers. Its edible fruits were once an important forage crop for sheep in mountainous regions of Southern Europe, as well as providing a feast for humans when puréed or roasted in the autumn.

fruits in the autumn and, if possible, showy leaves and bark.

Colourful shrubs also add interest in parks. Rhododendrons from the Himalayas, buddleias from China, lilacs, jasmines and forsythias ... the list is endless. Butterflies feast on the nectar from many of these whilst hoverflies and froghoppers can be seen on the flowers and leaves of the bushes. Moreover, the thickets of rhododendron offer refuge for the grey squirrels that comb the ground looking for fallen fruits from the forest trees.

FUNGUS FORAYS

Parklands make excellent places to study fungi. In fact, in some localities the parklands are *the* best places to look for fungi, over and above the purely natural habitats. Fungi thrive where there are plenty of plant remains in a damp environment, and in parkland there is often a great deal of dead wood which contributes to a rich pile of humus under the trees. In Western Europe the hedgerows, woodlands and parks blossom with edible and inedible fungi in the autumn. The damp September and October weather stimulates the fungi into reproductive activity and the larger species push up their 'fruiting bodies' – the most familiar part of the plant – in order to liberate their spores to the wind.

Most Europeans are keen fungus eaters, with the exception of the British who generally steer clear of any curious toadstool. The word toadstool is, in fact, just a general name for mushrooms and implies that they are inedible. There are many edible specialities to be found in woods and parks. Top of the list is the edible boletus – called variously the penny bun in England (because of its appearance), the *steinpilz* in Germany and the *cep* in France. Others include the morels, chanterelles, shaggy ink caps (which are very common on roadside verges), giant puffballs and woodland blewits. Beefsteak fungus – named after its rich red meaty colour – grows as a bracket fungus on parkland oaks and sweet chestnut trees.

Wood and parkland trees also harbour the coveted truffles that both the French and Italians are keen to mass-produce. These are fungi which grow in association with the roots of oak, poplar, hazel and hornbeam. Growing underground, they are seldom seen in their natural state, but when mature they are hard, black and rather warty in appearance. Their popularity as a gastronomic delight has led to suspensions of the truffle spores being injected directly into the roots of host trees in order to factory-farm them.

Research work in the United States indicates that small mammals of the woodland floor – and, in their case, chipmonks – are attracted to the smell of truffles, eat them and inadvertently carry their spores away with them to other areas of the wood. It is likely that small mammals in Europe carry out the same job, thereby spreading the

Above: *Shaggy pholiota is an impressive non-edible fungus which can be found growing from cracks in several species of deciduous tree, particularly beech. The same crack might produce a collection of these fruiting bodies for several autumns in a row. These are just the fruiting bodies of the fungus – its roots or mycelia ramify inside the tree trunk, progressively breaking it down.*

EDIBLE FUNGI

Wood blewit. *A woodland delicacy, these fungi can be found growing in gardens, beneath hedgerows and in oak woods.*

Edible boletus. *One of the most sought after fungi in Western Europe. It has pores on the underside of its cap instead of gills. There are many different species, most of which are edible.*

Chanterelle. *These edible orange fungi blossom from woodland floors. They smell of apricots and their gills are not properly formed.*

Horn of plenty. *This black, funnel-shaped fungus grows in the leaf litter of deciduous woods and is edible. It can be dried, ground and used for seasoning.*

Many different kinds of edible fungus can be found in woods and parks. They are looked upon as a delicacy in mainland Europe and compliment many dishes, but the British remain shy of their use and prefer to let them be eaten by insects or other fungi. Of course, there are some fungi which are extremely poisonous – the panther and death caps, for example, or the destroying angel.

Perhaps one of the most enjoyable things about fungi is their many and varied names. Apart from cauliflower, beefsteak and orange peel fungus, which are named on account of their appearance, there are those with such delightful titles as the amethyst deceiver, the earth star, the dryad's saddle and the wood woolly-foot! Not all of them are edible.

fungus to new trees.

Honey fungus is a particularly common parkland fungus. It is responsible for killing trees and can be recognised by its rich yellow-orange colour. It sometimes occurs on tree stumps, but is more often seen as plentiful brackets on living trunks. The damaging part of all fungi which live on and in trees is their network of filaments called mycelia. This runs through the tree, sapping its energy and feeding on its tissues. It is only when the fungi reproduce that they throw out a bracket on the outside of the tree – the familiar part that we see. If you see a bracket on a tree in your garden there is no use snapping it off and hoping that everything is all right. It's more than just the tip of the iceberg – that tree is fated.

Oyster fungus is another bracket fungus which can dispatch forest trees such as beech. Indeed many trees in Western Europe are killed by fungi. Storm damage to trees opens up wounds which are quickly colonized by airborne spores. The fate of the tree is then sealed. Fine beech trees are hollowed out and killed by oyster, honey and other bracket fungi, and silver birches often end their days thanks to the attractive white birch polypore fungus.

By destroying trees in this way, fungi do an excellent and essential job in woods and parks for

Above: *The large and conspicuous Dryad's saddle is a familiar sight in early autumn, growing from the stumps of deciduous trees. Sometimes reaching up to 60cms across, this fungus can also be found on living trees such as elm where it causes the wood to turn white. Like many other bracket fungi, it can be very destructive.*

Right: *The aptly named orange peel fungus can be found growing amongst the leaf litter or bare ground on tracks in parks, plantations and even gardens. Despite its title it is not edible.*

Above: *A familiar fungus of parks, gardens and commons, the fly agaric is normally associated with the silver birch and is nearly always found growing close to these trees. The muscarine poisons contained within its flesh give it hallucinogenic properties and once led to its use by certain tribes as part of their ritual festivities.*

other plants and animals. As decay progresses, the whole of the trunk may be hollowed out to provide a breeding site for owls, kestrels, woodpeckers and starlings. Moreover, all the wood and leaves that fall to the woodland floor are efficiently devoured by other fungi and turned into minerals. The woodland floor literally mushrooms with mushrooms; old tree stumps come alive with a variety of fungi all working to reduce the remaining wood to a pulp. Other wildlife moves in – numerous small invertebrates scurry amongst the leaf litter and conceal themselves beneath the bark, and mosses, liverworts and ferns provide them with shelter.

PARK BIRDS

Parks, like gardens, are good places to study birds. The open areas of grassland, clumps of trees, thickets and formal flower beds all harbour different types of bird. You can walk through a park with a pair of binoculars and get a good look at many birds without worrying them. Another advantage is that many parkland birds are tame; some may be regularly hand-fed, and most are used to people walking, sitting and playing nearby. Most common garden birds are regular

TWENTY COMMON FUNGI	
Beefsteak (Ox-tongue)	Edible but not tasty
Birch polypore	'Razorstrop fungus' – once used for stropping razors
Blewits	Edible and delicious
Cauliflower	Edible; fiddly to prepare
Cramp balls (King Alfred's cakes)	Found on ash stumps
Death cap	Fatal
Earthball	Inedible and could be confused with puffballs
Fly agaric	Usually found under birch trees
Green wood cup	Makes oak twigs on woodland floor go green; was once used as green inlay on Tonbridge-ware
Horse mushroom	Found in parks and orchards
Horse's hoof	A bracket fungus; rare in Britain; inedible
Jew's ear	Found hanging from dead elders; edible
Orange peel	Sometimes common on paths; inedible
Parasol, several species	Edible and delicious; grows in rings
Penny bun	Edible and delicious
Puffball, several species	All British puffballs are edible; cut slices and fry
Slippery Jack	Looks wet, feels slippery
Stinkhorn	Often smelt but rarely seen
Sulphur tuft	Groups seen growing from cut wood, often under pines

Above: *Hopping back and forth across the open space of a park lawn, the striking magpie can be easily recognised by its black and white plumage and long tail. A great opportunist, this relative of the rooks and crows can be seen scavenging in rubbish bins for scraps, hoarding surplus food as well as colourful and shiny objects in its untidy nest.*

visitors to parks – blackbirds, robins, sparrows in their hundreds, blue and great tits as well as larger species such as crows and jackdaws.

The large areas of tidy turf so often found in parks offer an ideal feeding area for flocks of town birds. It is as if this grassland habitat has been especially created by man to attract them. Regular mowing maintains a short turf where food is easy to find, particularly wireworms, earthworms and leatherjackets – favourite foods of many species. In coastal towns the grassy spaces are used by seabirds which sit out the high tide in comparative luxury. When the high water covers their mudflat feeding grounds, the birds fly progressively inland to rest until the turn of the tide. In the wild, marshes are used for the same purpose but where man lives close to the sea his green pastures make a good alternative. The seabirds find food there too. Black-headed gulls, herring gulls and oystercatchers can often be seen occupying playing fields, recreation fields and parks in great numbers. They too search for earthworms, and other denizens of the soil, and frequently position themselves midway between housing, people and roads – a white mass of birds in a sea of green.

Regular visitors to large fields and parks are starlings. They alight and fan out, thoroughly combing the ground for insects. Woodpigeons are parkland birds too. They construct nests along the horizontal branches of trees and can be seen eating acorns and beech nuts under the parkland trees. Masses of woodpigeons often roost communally near large towns and cities. Thousands cover the trees so that sometimes the canopy of leaves and branches appears blue instead of the usual green. This bird has made great adaptations to man-the-farmer and is a pest in both field and forest – eating seeds and crops. It has also adopted gardens and parks as alternative feeding grounds.

Autumn visitors to parks are fieldfares and

redwings. These members of the thrush family breed in Northern Europe during the summer and move southwards for the winter. The fieldfare is the larger of the two, with chestnut on its wings and back. The redwing has red on its underwings and flank. Both birds eat fruits and insects – the berries of hawthorn, mountain ash (rowan), yew and holly, as well as worms, spiders and snails.

Two very similar visitors to parks and gardens are the mistle and song thrushes. Both handsome birds with spotted breasts, they could be confused with fieldfares. Both birds eat fruits and seeds, insects and their larvae, as well as snails and worms. The song thrush is well known for its habit of using a particular stone as an 'anvil' on which to smash snail shells.

Lapwings are often seen in parks. These curious birds are also called peewits after their familiar sound, and are easily recognised by the long crest on their heads and their peculiar stooping flight. During the winter they gather in large groups and feed on wireworms and leatherjackets. They can be seen on most open stretches of land: parks, fields, moorland and coastal marshes and, most particularly, farmland. They are beneficial birds,

Above: *The jackdaw is another opportunist bird which is totally at home in the urban environment, scavenging from bins and coming close to humans in order to take food. In the wild it will steal food from other birds and rob their nests of eggs and chicks. A fearless thief, this bird will even perch on the backs of cows and horses, plucking out their fur to line its nest.*

Left: *Like heronries, the communal nesting sites of rooks, known as rookeries, often have historical roots stretching back over several centuries. One of the largest rookeries ever known contained over 6,000 pairs, their nests silhouetted on the skyline in spring. Where their supporting trees, such as elms have died out, the colonies move on to other trees or more permanent structures such as pylons.*

Above: *Pheasants may occasionally stray into the rougher areas of parkland from adjacent fields, commons and copses where they are bred or live wild. These birds are frequently seen in pairs, the cock being easily identified by his red eye patch and brighter plumage.*

eating many pests such as wireworms and leatherjackets. In winter they may be found near water – on freshwater margins and coastal estuaries.

Parklands and playing fields would not be complete without the sound of the skylark melodiously singing high above the ground. This bird of open spaces sometimes reaches heights of 300 metres when, no more than a speck in the sky above, its shrill song may still be heard. When on the ground, these charming birds are very secretive about their nest sites. Often sited in a small hollow in the ground, the nest is built of grass and hair – usually amongst tall grasses in rough areas where it is well concealed. Often they will allow you to approach to within several metres before flying off.

Ancient park trees offer great potential for three common woodpeckers. The green woodpecker is the largest of the three. It sometimes comes down to the ground and will spend a considerable length of time probing the turf for insects with its long bill. It also has an extremely long and sticky tongue which it uses to pick up ants. The green woodpecker doesn't make a pecking sound, unlike the greater spotted species which makes very audible bursts of about 1 second in length. It can be distinguished from the lesser spotted woodpecker which makes bursts lasting 2 seconds. Both birds make about 15 strikes of the bill against the wood per second, rather like a pneumatic drill. Since this is enough to cause concussion, it is not surprising to learn that woodpeckers' skulls are specially cushioned to protect their brains from this way of life! All three have moved into suburbia and can be found in parks where food and nesting sites are plentiful.

Apart from other wild birds such as whitethroats, stonechats and magpies, pet birds are well established in some European parks. Budgerigars and parakeets, escapees from homes or birds deliberately released by their owners, find the warmth of city life and the suitability of habitat to their liking. Ring-necked parakeets, originally from Africa, are now well established in large flocks on the Isle of Thanet in north-east Kent and in some South London parks.

PARK BIRDS	
Dunnock	Mistle thrush
Fieldfare	Nuthatch
Goldfinch	Owl
Great spotted woodpecker	Redpoll
Greenfinch	Redwing
Green woodpecker	Siskin
Kestrel	Song thrush
Lapwing	Sparrowhawk
Lesser spotted woodpecker	Treecreeper
Linnet	Tree sparrow

THE PARKLAND HABITAT

Urban parks, rather than national parks, offer an ideal habitat for wildlife close to towns. Ornamental trees, stumps, pools and flower-rich ancient pastures play host to a rich variety of insects and fungi. The large open spaces provide ample room for movement for mammals within their home ranges and for migrant birds to make passing visits. Neat lawns may be combed by common bird species such as starlings and blackbirds as they search for small invertebrates hiding amongst the short turf, and showy flower beds make available a rich source of nectar and protein-rich pollen.

In the wilder areas, wildlife is given the chance to thrive without the persistent application of herbicides, insecticides and fungicides – all substances that have had a detrimental affect elsewhere in the urban environment.

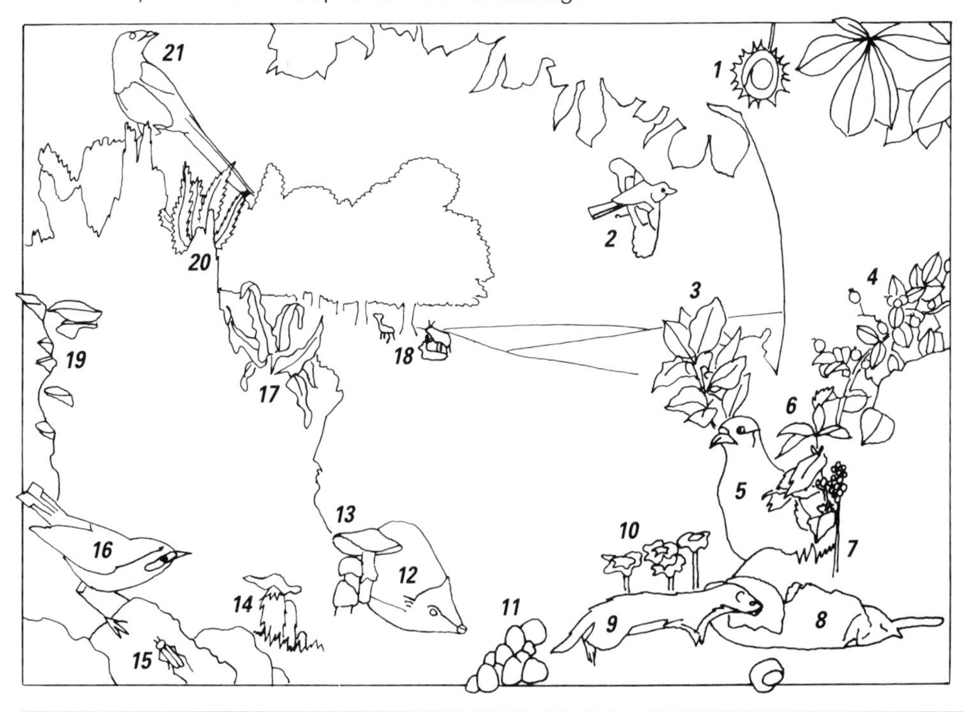

***1** Horse chestnut*
***2** Jay with acorn*
***3** Deadly nightshade*
***4** Rose hips*
***5** Cock pheasant*
***6** White deadnettle*
***7** Lords-and-ladies (fruit)*
***8** Rabbit*
***9** Weasel*
***10** Sulphur tuft fungus (old)*
***11** Sulphur tuft fungus (new)*
***12** Hedgehog*
***13** Panther cap fungus*
***14** Shaggy ink cap fungus*
***15** Wasp beetle*
***16** Wheatear*
***17** Hart's tongue fern*
***18** Red deer*
***19** Ganoderma fungus*
***20** Hard fern*
***21** Magpie*

PARK INSECTS

Butterflies and moths have always been found in parks – but now that the public has become more aware of them, strenuous efforts are being made to create butterfly gardens to ensure their continued presence. Butterfly gardens must cater for two things – a source of nectar for the butterflies and the correct food plants for their caterpillars. Usually nectar plants and caterpillar food plants are completely different.

Herbaceous borders in parks are rich in nectar sources. Many popular plants yield liberal amounts of nectar – lavender, Michaelmas daisies, verbenas, buddleias, ice-plants (sedums), salvias, marigolds and zinneas to name but a few. Beds containing such species will be rewarded with visits from the aristocratic butterflies: red admirals, small tortoiseshells, painted ladies and commas, as well as small and large whites. Areas where nettles, wild cabbage, thistles and other plants grow will provide food for their larvae. Corners where wild grasses, docks and other plants are left to thrive will also serve to attract the small and large skipper, small copper, small

***Above:** The great banded grayling is a central and southern European species often found in woods and parkland where it will fly up and down shady paths, alighting on tree trunks where its wing pattern helps to keep it well camouflaged. It is sometimes attracted to the flower beds in parks and gardens where it feeds on marigolds or petunias. The butterfly does not occur in Britain. It is the largest member of the brown family of butterflies in Europe. Its caterpillars are longitudinally striped black, white and orange. They feed on various grasses including rye and brome.*

heath, meadow brown, gatekeeper and occasionally the secretive ringlet butterflies. The long grass is ideal as a food plant for the caterpillars of the skippers and browns.

Parks are also ideal places to find moths. These insects are just as important as butterflies but tend to be neglected because most do not have bright colours. In fact, there are plenty more species of moth than there are butterflies – at least 20 times more. Parkland trees and tree-lined urban roads used to be a favourite haunt of the moth-hunters several decades ago. They used a method of attracting moths called 'sugaring' which is still occasionally used today. With this, a rich sugary mixture was pasted onto tree trunks just before dusk. The moths would smell the sweet mixture and come to feed, thereby allowing entomologists to study them.

You don't need to go to this extreme in order to get a close look at many species of moth however. Look on tree trunks anyway, since some moths conceal themselves on the bark during the day. Look, too, under windowsills, under eaves and on wooden palings. Moths can even be found in public conveniences where lights are kept on all the time. In one recently I found several species of moth and over 50 hibernating lacewings!

In urban areas, the soot and grime that collects on trees has meant that those moths that normally rest (and are camouflaged) on light-coloured tree trunks now stand out obviously to predators. However, many have evolved a darker form which allows them to blend in with the dirty background. One moth is often singled out as demonstrating this 'industrial melanism' – the peppered moth. Normally a brown and white speckled insect that camouflages well against the bark of birch trees, its melanic form varies from dark brown to black.

Above: *The bright colours of the lackey moth caterpillar are in direct contrast to the dull brown of the resulting adult which camouflages well against tree trunks and is less likely to be noticed. The larvae can be found feeding on rose, bramble and a variety of trees.*

COMMON PARK MOTHS	
Angle shades	Lime hawk
Blair's shoulder knot	Magpie
Brindled beauty	Muslin
Brown tail	Oak hook tip
Buff ermine	Peppered
Buff-tip	Poplar hawk
Burnished brass	Privet hawk
Cinnabar	Puss
Common swift	Setaceous Hebrew character
Dotted border	Silver Y
Elephant hawk	Swallow-tailed
Eyed hawk	Swallow prominent
Garden carpet	Swallow kitten
Garden tiger	White ermine
Heart and dart	Winter

MOTH OR BUTTERFLY?

Butterflies and moths were not evolved to fit neatly into little classification boxes, and there are so many exceptions to each rule that the excerise is often not worthwhile! However, there are a few general guidlines that can be borne in mind when trying to identify these insects.

Generally, butterflies fly by day, have bright colours, rest with their wings clapped together, have clubbed antennae and are not so hairy as moths – characterisitics exhibited by the orange tip butterfly below.

Moths generally fly at night, rest with their wings flat over their 'backs', have drab colours and anything but clubbed antennae, and are usually very hairy. They also have little hooks on the rear edge of their fore wings which allow them to link their fore and hind wings together. Butterflies do not have this. The puss moth below shows some of these characteristics.

Orange tip butterfly *(male)*

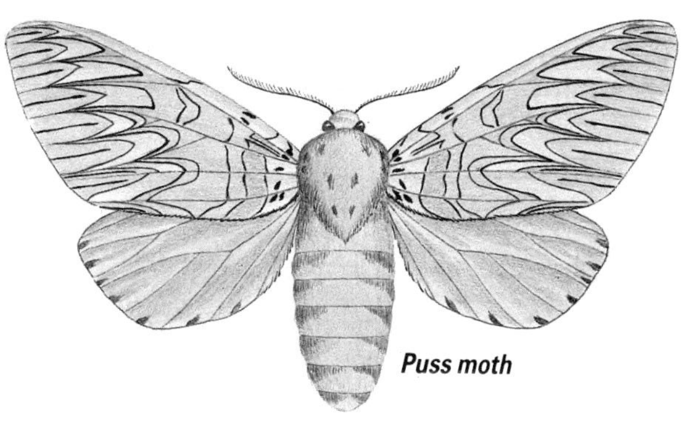

Puss moth

Honey-bees and bumble bees are other insects regularly seen working the flowers in parks. Salvias, foxgloves, fuchsias, heathers, cotoneaster, marigolds, heliopsis, skimmia and grape hyacinth are some of the regular garden plants they visit. Magnolias, willows and ivy also arrest their attention. On the grassy lawns dandelions and clovers are good nectar and pollen sources and in the rougher areas comfrey, escaped raspberry and mallows are never missed.

Apart from providing further feeding grounds for urban bees, large open parks and playing fields also serve as places where drones – male honey-bees – have their assembly areas. You may sometimes be aware of the perpetual hum of bees on a fine day, yet you cannot place the sound. This is because they will all be way above you – out of sight perhaps 100 metres above the ground. This happens through the spring and summer when a new queen is needed for the hive, and begins with the drones from several hives gathering together above a park or similar open space.

The purpose of this behaviour is to get one of the hive's virgin queens mated. Once out of the hive on her nuptial flight, the queen – it is thought – can hear the drone group and makes her way to it. Once there, she is pursued by the eager males bent on mating with her. One of them gets the honour and mates with the queen in full flight. She then returns to the hive with his torn off genitalia hanging out of her abdomen. After a few days she starts to lay the eggs of the next generation. Having contributed his fleeting service, the unlucky male is left to die.

***Above:** The impressive silver-washed fritillary can be attracted into flower beds in parks and gardens adjacent to wooded areas. It is commonly found in woodland clearings where the wild violets on which its caterpillars feed can still be found.*

A GUIDE TO SOME PARKLAND MOTHS

The rough pastures of parklands are likely to support plenty of different types of day-flying moth. They are at the mercy of predatory birds, lizards and small mammals so, unlike their nocturnal relatives, many have evolved bright colours like those of butterflies and have made their bodies unpalatable by storing plant toxins passed to them from their larval stage. Birds associate the nasty taste with the colour pattern and do not attempt to eat others with the same coloration.

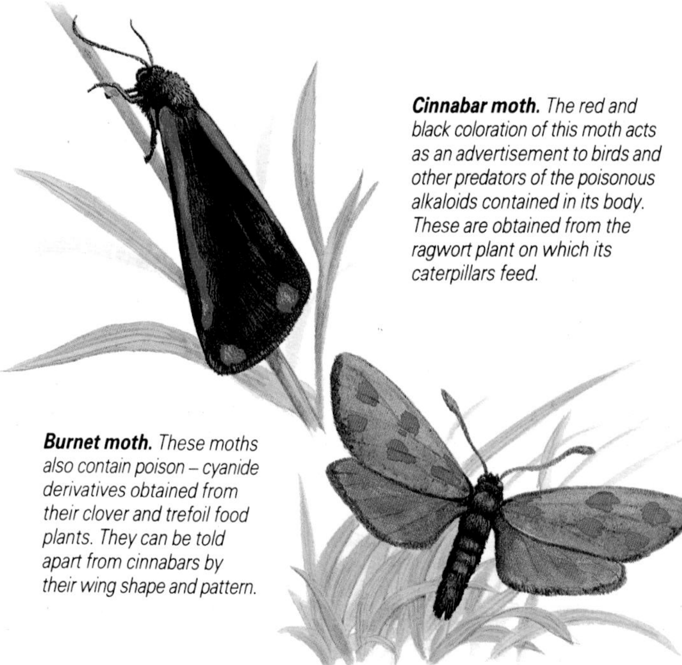

***Cinnabar moth.** The red and black coloration of this moth acts as an advertisement to birds and other predators of the poisonous alkaloids contained in its body. These are obtained from the ragwort plant on which its caterpillars feed.*

***Burnet moth.** These moths also contain poison – cyanide derivatives obtained from their clover and trefoil food plants. They can be told apart from cinnabars by their wing shape and pattern.*

Above: *Wild pansies, known commonly as heartsease, are eaten by the caterpillars of the fritillary butterflies. They grow wild throughout Europe and cultivated varieties are common in the flower beds of parks and gardens alike.*

Left: *The painted lady is a strong migrant found throughout Europe. Its caterpillars feed on thistles and nettles, whilst the adults can be seen taking nectar from knapweeds, Michaelmas daisies, buddleia (shown here) and other plants. The females lay their eggs singly on the food plants and the tiny black larvae grow and feed together inside silken webs amongst the leaves.*

One of the highlights of a naturalist's visit to a park is the discovery of an old tree trunk or stump. If it is positioned under other trees, it may be a mass of vegetation – ferns, mosses and liverworts are particular colonizers here – and if the bark is damp it can be peeled back to reveal a whole variety of creatures. Such places are complete habitats by themselves – enough to satisfy any bug-hunter, botanist or beetle freak!

A whole day spent exploring the wildlife of tree trunks would be well rewarded – woodlice, centipedes, millipedes, tiny and gigantic beetles, as well as slugs and snails can all be found lurking in damp cracks and crevices. Woodlice are perhaps some of the most ubiquitous inhabitants of dead wood and other damp, dark places. Although land-living crustaceans, their closest

PARK BUTTERFLIES	
Brimstone	Meadow brown
Common blue	Red admiral
Gatekeeper	Small copper
Large skipper	Small white
Marbled white	Speckled wood

Emperor moth. *This moth may be found where there are large patches of heather on which its caterpillars feed. The sexes are not alike. The male, shown here, is smaller and darker than the female and has large pectinated antennae.*

Speckled yellow. *A familiar sight in woods and rides, the bright colours of this moth could be confused with those of a small butterfly. Its eggs are laid on wood sage, woundworts and dead-nettles.*

Garden tiger. *These moths are also poisonous, containing various active chemicals which affect mammalian nervous systems. Sparrows will often disturb these bright moths and attempt to eat them until they discover their toxic content.*

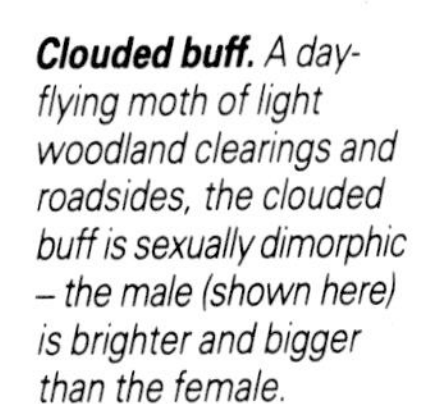

Clouded buff. *A day-flying moth of light woodland clearings and roadsides, the clouded buff is sexually dimorphic – the male (shown here) is brighter and bigger than the female.*

Oak eggar. *The males of this species are often seen flying frantically up and down hedgerows or clearings. Their rich colour and rapid flight are very distinctive. The females are only seen occasionally.*

living relatives are crabs and shrimps. As if to emphasize this link, they still need to live in a moist environment. Many different species can be found, some of which are capable of rolling up into a tight, shiny ball. They were once swallowed live as a supposed remedy for stomach upsets, hence their alternative name of pill-bug!

If you roll over such a log to examine its smaller inhabitants, do remember to roll it back again – not only for the sake of the wildlife but also for the next visiting naturalist. Furthermore, it is not a good idea to go around stripping bark from trees in order to find the beetles and their larvae that might be lurking there. There are plenty of trees which already have peeling bark which you can peep behind.

Spiders are very numerous in parks and gardens. Sometimes the grass looks as though it is covered completely with silk, especially during autumn. This gossamer is produced by a group called the ballooning spiders. They are tiny creatures which employ a novel method of hang-gliding in order to get about. First, they firmly attach themselves to a leaf, stick their abdomens into the air and produce many fine strands of silk

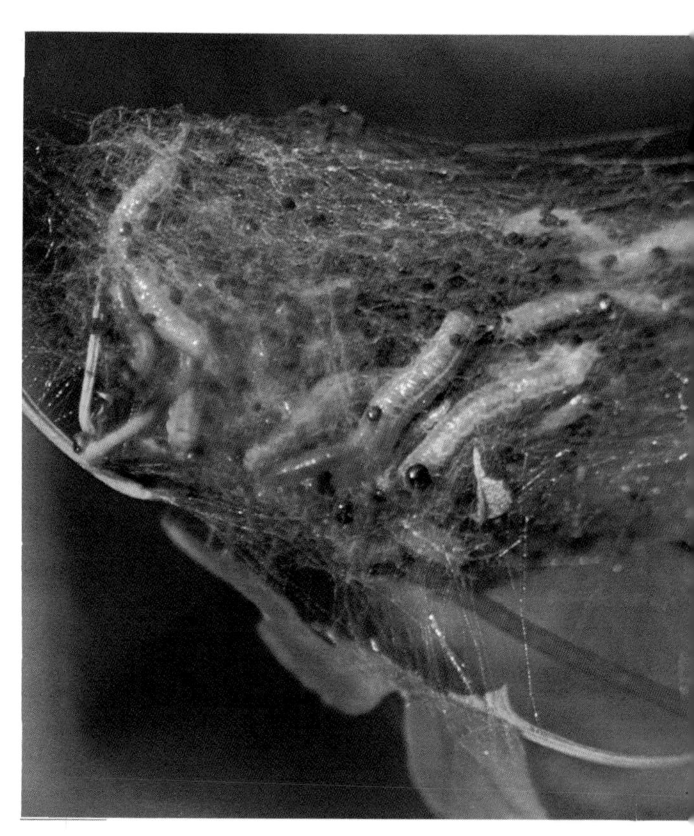

TREE STUMP CREATURES

Beetles, larvae	Pale cream and orange; difficult to identify
Centipedes	Common predators of insects and spiders
Earwigs	Wings rarely used; females exhibit parental care
Flea beetles, several species	Tiny jumping beetles
Ground beetles	Black or blue; beneficial insects
Harvestman	Has eight legs; body in two segments
Lacewings	Sometimes found hibernating under bark
Moth chrysalises	Several species found under bark – tiger moths, noctuids
Moths	Bark is daytime resting place for many moths – angle shades, carpet moths, Merveille du Jour
Millipedes	Common predators of other invertebrates
Pseudoscorpions	Tiny scorpion-like creatures 3 mm long
Scorpions	Southern Europe under bark and stones.
Snail's eggs	White, glutinous; in groups where damp
Spiders	Many living on or under bark
Springtails	Tiny primitive insects which jump
Stag beetles	Larvae bore large holes in wood
Woodworm	Tiny tell-tale holes in bark
Wood ants	Scavenge over wood for food

Above: *The Argiope spider belongs to the same family as the common garden spider but is not so readily seen. It belongs to a large group known as the orb web spiders and sets its wheel-shaped net across vegetation in rough grassland to catch its prey of flies and other insects. The web can be distinguished from those belonging to other spiders since this species has developed the curious habit of marking its web – a personal signature in the form of a zig-zag design. In good localities, such as the flowery undergrowth at the side of streams, they may be very common; perhaps two every metre or so.*

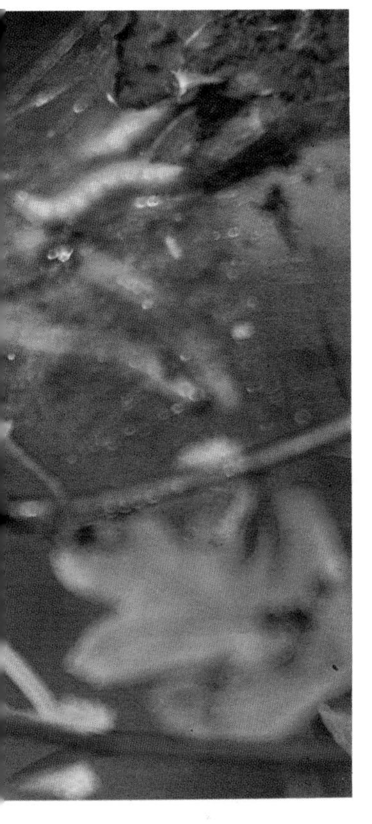

Above: *The praying mantis has highly modified fore legs armed with long spines with which to grip its prey. Its stick-like green body enables it to lurk unseen amongst vegetation, waiting motionless for another insect to come within striking distance.*

Left: *These sawfly larvae are likely to be found in hedgerows around gardens and parks. They spin silken tents over their feeding area and move on to new twigs only when they have stripped an area bare of leaves.*

which are taken away by the wind. Then, at the appropriate moment, they let go of their leaf and off they go – kiting across the country. The caps of footballers' boots become covered with this fine white gossamer as do the eyes of running dogs. In fact, in the last war the silk was even mistaken for some sort of chemical warfare! It has proved, however, to be a very effective means of dispersal, carrying the spiders to new feeding and breeding sites. The great naturalist Charles Darwin noted that the rigging of his boat, *HMS Beagle*, was covered with such silk 100 km from land.

Spiders may be discovered along walls and palings bordering parks, particularly the garden and *Argiope* spiders. Looks elsewhere for them, too, on the branches of trees, on heathy areas, sandy areas and in flower beds. The crab spiders are often found lurking amongst the petals, lying in wait for flies. Some of them are quite gaily coloured, either pink, white, yellow, black or grey.

PARK INVERTEBRATES	
Beautiful damselfly	Glow-worm beetle
Cicada	Harvestman
Common field grasshopper	Horsefly
Common green capsid bug	Lacewing
Common wasp	Scorpion fly
Dor beetle	Soldier beetle
Earwig	Stag beetle
Emperor dragonfly	Tiger beetle
German wasp	Wood ant
Giant wood wasp	Wood cricket

If you see a whole host of butterflies' wings beneath a thistle or knapweed plant, look carefully at the foliage since you will probably find a praying mantis concealed within the flowers. This carnivorous insect, named after its typical praying posture, occurs throughout Southern Europe. It uses its disguise to catch unsuspecting insects which are attracted to the flowers. Crab spiders often employ the same strategy.

Two other interesting insects may also be found in South European parks. These are the ant lions and the ascalaphids. Ant lions are rather like unco-ordinated dragonflies but are nocturnal and are frequently attracted to lights. The presence of their larvae in parks is seen in sandy areas by a little circular depression in the sand. The carnivorous larvae lie in wait at the base of this sand funnel and, as soon as a little insect falls into the depression, it descends into the jaws of the waiting larva.

Ascalaphids have stout bodies and long clubbed antennae. They are carnivorous like dragonflies and fly about over rough grassland, catching flies and other insects. They are the giant relatives, along with ant lions, of the lacewing order of insects. On dull, drizzly days they are sometimes the only insects that can be found flying. It would seem that they have great adaptations to the British climate but they are not found further north than France.

PARK MAMMALS

Deer are the largest and most obvious mammals to be found in urban parks, many being the descendents of deer once kept in royal parks. The commonest species is, without doubt, the fallow deer, believed to have been introduced to Northern Europe by the Romans. Red deer, a taller species native to Northern Europe, are also occasionally seen in parks. These handsome forest mammals have been gaining in popularity in recent years due to the rise in deer farms. The smaller roe and sika deer may sometimes be seen, and even reindeer from the Arctic or lamas are loose in some European parks – a reflection of the ease with which unfamiliar animals can be readily moved around the world.

Fallow deer survive well in parks since they feed on the grass and derive cover from the bracken which is so often found there. The large areas of open land provide ample space over which they can roam and establish territories, and people are always generous with food hand-outs. The only problem is that, through inter-breeding, colour mutations occur frequently. White forms of fallow deer – which normally sport a bright chestnut coat with white spots in summer,

turning to a plain grey brown in winter – are common, as are the black or melanic forms. There must have been many such white stags about in parks long ago to warrant so many English pubs called The White Hart!

Like male red deer, fallow bucks sport impressive antlers. These bony growths are shed yearly, each new pair growing progressively bigger and more complex every year. During the spring and summer growth period, the antlers are covered in thickly-haired skin known as velvet. This is rubbed off by the stag when the antlers are fully grown and can sometimes be found on tree trunks. It is prized in the East as an aphrodisiac.

The antlers are used for fighting during the mating season and also for spreading scent from the facial glands to mark territory. The antlers are cast aside by the following spring and may occasionally be found amongst the undergrowth. They are sometimes chewed by the deer themselves for the rich minerals they contain.

Parks with plenty of undergrowth will undoubtedly attract hedgehogs. They roam about feeding on invertebrates and frequently enter gardens adjacent to parks. They may find sanctuary in bramble thickets, rough and wild corners and even dense flower beds. A wide range of small mammals – field voles, yellow-necked field mice and shrews – will also live in the long grass, bramble thickets and scrub, thriving on the many seeds, insects and other invertebrates found there. In turn, they attract foxes which make their earths in dense undergrowth.

Less commonly, the setts of badgers may be found in the wilder corners of parks. Signs of their presence are quite common – sett entrances,

Above: *Fallow deer are commonly seen in deer parks where they often become tame enough to feed from the hand. The bucks produce new antlers each year which are used to assert their status within the herd and are cast off the following spring or summer. The development of antlers depletes the animal's reserves of calcium so the discarded growths are often chewed in an effort to replace the lost mineral.*

PARK MAMMALS	
Badger	Field vole
Bank vole	Fox
Common shrew	Mole
Grey squirrel	Rabbit
Fallow deer	Wood mouse

Above: *The rabbit is regarded as a native of north-west Europe but was introduced by the Normans and farmed in organised warrens for fur and meat. These creatures suffered dramatically with myxomatosis but an outdoor strain of scrub rabbits avoided the flea-ridden burrows and, therefore, the disease and were able to build up populations once again. Myxomatosis strikes periodically but the rabbit is so resourceful that it will remain a familiar sight in parks, on commons and in other open spaces.*

footprints, droppings and grey hairs caught on wire fencing are just a few things to look for.

A much more obvious inhabitant of parkland is the rabbit. These familiar creatures can often be seen during the day, and their prolific droppings are a familiar sight on grassland. They eat a wide variety of vegetable food and obtain maximum benefit from it by double digestion. At night the liquid droppings – as a form of slurry – are evacuated into the mouth to retrace their steps through the animal's alimentary canal. Vitally important bacteria then extract further nutrients from the food which is finally passed out in the familiar form of hard pellets.

Much parkland is ideal for rabbits, allowing them access to well-drained soil in which to make their extensive burrows. One large warren may contain several hundred rabbits, with each dominant buck having his own territory. They seldom feed further than 140 metres from their warrens and, whilst grazing, frequently sit up on their hind legs to check for signs of danger. One rabbit will alert its neighbours to the presence of an enemy by thumping the ground with a hind foot before scurrying to safety. The flashing of their white tails whilst they do so also acts as a warning.

Apart from man, the rabbit's chief enemies are the fox, weasel and the larger owls and hawks. These creatures, too, may occasionally be seen in parks during the day, though most confine their hunting activities to the hours of darkness.

Hollow trees in parks make ideal winter roosts for another parkland mammal – the bat. The overall problem in Europe is that there are now not enough roost sites such as hollow trees, so those that occur in parks offer an important refuge for this group of interesting insectivores. The species most likely to roost in parkland trees are noctules, Leisler's barbastelle, and Natterer's bats, and occasionally pipistrelles.

Where bat boxes have been erected (see page 153), they may be used for hibernation or rearing young. Chattering and squeaking sounds coming from the box are a sure sign that it is being used as a roost. On some occasions, however, such boxes attract other parkland creatures. Tawny owls and woodpeckers interfere with the bats, and treecreepers nest where the bats should go.

Grey squirrels – those North American imports – are a pest in some urban parks in Britain. Although nice to look at, they can cause serious damage – stripping trees of their bark and digging up the soil to find or store food.

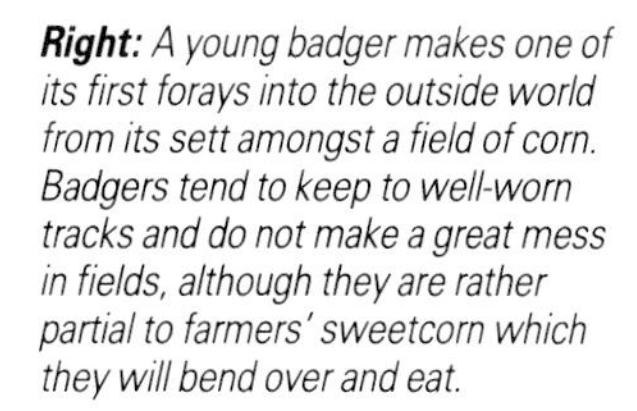

Right: *A young badger makes one of its first forays into the outside world from its sett amongst a field of corn. Badgers tend to keep to well-worn tracks and do not make a great mess in fields, although they are rather partial to farmers' sweetcorn which they will bend over and eat.*

PARK WILD PLANTS

It might be that you walk through a park regularly each day – perhaps with the dog, perhaps to work, or perhaps just for relaxation. Whatever your motive, you are sure to see a lot of wild plants in flower and, unless you are a very keen botanist, it's unlikely that you will know all the plants you see. Parks contain all types of wild flower – from the common to the unusual. Try to spot a new one each day and identify it – you may eventually find well over a hundred species.

If your walk takes you along a river bank there may be clumps of comfrey or Himalayan balsam growing abundantly. Betony and field woundwort add purple colour to damp banks, and golden saxifrage brightens up muddy areas. Old parks often have boundary ditches and these, if overgrown and out of the direct rays of the sun, may well be blossoming with figworts, water dropwort, irises, mosses and liverworts.

Rough areas of long grass can be very rewarding; hound's tongue, deadly nightshade and crane's bill are surprises to look out for. Shady areas near hedgerows may harbour plenty of wood sage – a great attraction for honey-bees in the autumn, willowherbs, wood spurge, sun spurge and Goldilocks buttercup. You may also find tall stands of purple foxglove contrasting with the bright yellow of creeping buttercups. Rare parkland plants to find are thorn apple – often where there has been disturbed ground – and henbane. Both are poisonous.

In the spring the parkland grass may be studded with primroses or cowslips. Hedge banks may have red campion, stitchwort, bluebells, violets and, if you are lucky, the delicate town hall clock or moschatel.

The summer brings lots of white and yellow umbellifers to hedgerows and clearings: cow parsley, hogweed and dropwort. There may also be alexanders (near the coast), fennel or ground elder (a real menace in the garden), and pignut or yarrow. The grass may be full of yellow vetchling, silverweed and St John's wort. On a hot sunny day listen out for the melodious notes of the yellowhammer and skylark as well as the popping of exploding gorse pods. Rougher areas of scrub may have bushes of hawthorn, sloe (blackthorn), wild privet, elder, field maple and oak. Hairstreak butterflies can be found playing around these bushes in the summer.

Parkland hedgerows are rich in fruits during autumn. White bryony has poisonous red fruits which hang in cascades from branches and even up the wires of telephone poles, as do wild hops and convolvulus. This bryony used to be called the mandrake and is meant to make a terrible screaming sound when pulled from the ground.

PARK WILD FLOWERS	
Brambles	Harebell
Clovers	Heather
Crucifers (most)	Mallows
Daisy	Marjoram
Dandelion	Mayweeds
Deadly nightshade	Mints
Dog's mercury	Old man's beard
Dog rose	Privet
Dog violet	Scabious
Fleabane	Thistles
Foxglove	White bryony

Above: *The yew is a dioecious tree, its male and female blossoms being born on separate trees. The poisonous female seed is surrounded by a protective red wax-like cup – a favourite (and edible food) of many birds and small mammals in the autumn.*

Left: *The wild grasssland of some parks supports a rich variety of wild flowers, including the ox-eye daisies, hawkbits and common daisies seen here. If the area has never been ploughed or dosed in organic herbicides then a rich flora will result. In time the airbourne seeds of orchids such as the early purple, common spotted and bee orchids will germinate to grow alongside other relic species such as adder's tongue and bedstraws.*

Its roots are thought to aid conception, and have magical powers.

Other colourful fruits are the hawthorn, elder and holly – all of which provide an autumn feast for birds such as the fieldfare and redwing. Many of the trees and shrubs also produce nuts at this time of year – varying from the shiny brown conker to the curiously shaped beech nut. Honeysuckles brighten many hedgerow banks with their clusters of red fruits, whilst the blackberry is a common sight almost everywhere.

Grass in parks will often be worn down very short where people walk, even breaking out into bare patches. Even here, it is surprising to see just how many plants thrive under these adverse conditions. Those that do tend to be miniaturised, a situation which also occurs on cliffs exposed to salt spray and high winds. Look for birds' foot trefoil, early purple or common spotted orchids, clovers and speedwells. Yellow rattle – an indicator of rich meadows – and lousewort often grow by the side of grass tracks or in the middle of footpaths. Similarly, orchids push up through footpaths even though there is lots of space elsewhere. Further headaches for wardens!

IN THE ROUGH

The wilder parts of parks, commons and golf courses harbour much interesting wildlife. If nature is left to take its course, a rich grassland habitat may develop where many species of wild plant can be found. Insects, birds and small mammals enjoy the freedom of these wild open spaces. The grass may be full of the ant hills of the yellow ant – these can be so numerous sometimes that it is impossible to walk in a straight line. Heath spotted orchids, helleborines and butterfly orchids may occur where it is lightly shaded. Lizards, slow worms, numerous beetles, snails and grasshoppers may frequent the pastures. Ground nesting birds like partridge, pheasant, skylark and meadow pipit find secluded spots to rest and feed and butterflies like skippers, meadow browns, ringlets and coppers frolic in the sunshine.

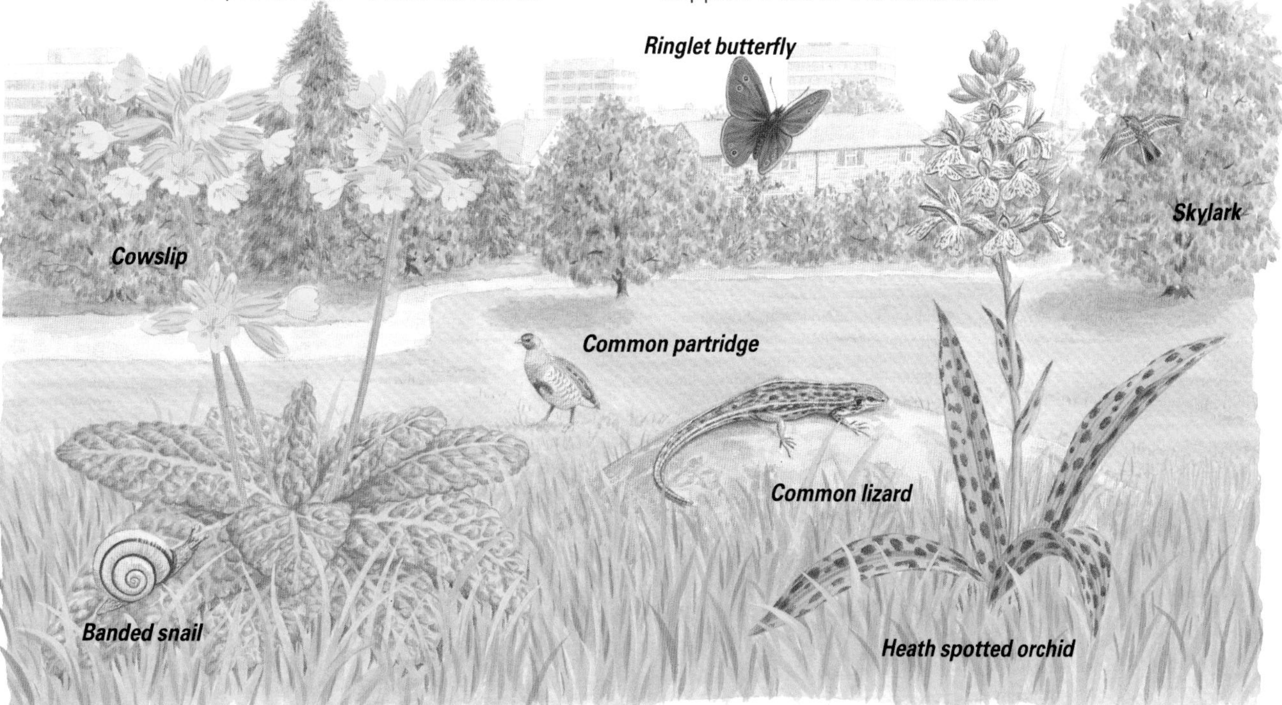

There are an infinite number of things to discover along the edges of roads and railways. We spend so much time travelling through the countryside that we are bound to see many of the more obvious and common forms of wildlife – bright splashes of wild flowers along the verges, hovering hawks and bounding mammals – but now that so much of the countryside has become impoverished, the road and railsides have become havens for many rare species as well. Such habitats are the most likely places to look for some species, the giant hogweed shown here being one example.

ROADS & RAILWAYS

Roads and railways both act as reservoirs for wildlife. Like gardens, they are refuges for plants and animals that have been driven out from surrounding agricultural land, urban developments or industrial areas. Contrary to opinion, the green corridors along the edges of man's well worn highways are rich in flora and fauna alike; they thrive in the total absence of man and his harmful chemicals that might be found elsewhere in their environment. Life in the middle of a busy motorway or along the edge of a mainline track is not all a bed of roses however – only the fittest and most able species can survive in the forgotten pastures of the verge, and not all have the ability to adapt and make the most of these new man-made habitats.

These 'linear nature reserves' may well seem to us to offer a somewhat oppressive environment for wildlife, but in many cases the opposite is true. Without interference from man, populations of plants and animals quickly build up. Furthermore, the chemical content of the soil in such places may be far more beneficial to certain plants than that of the surrounding agricultural land. There is still pollution to contend with but this has not always proved the disastrous problem that many of us think (see page 117).

Another reason for the success of these linear reserves is that they provide a continuous corridor for the distribution and spread of the wildlife found there. In many cases, the grassy habitat of verge or embankment stretches for many kilometres, allowing plants and animals to colonize in both directions by their own powers of distribution. Finally, food is often in plentiful supply, particularly along the motorway or roadside. The insect victims of passing traffic, for example, provide a ready feast for ants and various insectivorous birds, whilst larger creatures – the inevitable hedgehog and squirrel – are taken as carrion by bigger birds such as magpies and carrion crows.

HITCHING A RIDE

The probability of seeds being moved from one place to another, even from one country to another, by cars, trucks and trains is very high indeed. It is easy to imagine the seeds of grasses and other wild plants becoming stuck in mud under bumpers and in tyre treads only to be washed out many miles later in the next rainstorm. The movement of juggernauts across Western Europe is likely to spread all sorts of interesting seeds, and the fruit lorries which ply daily between Spain, Italy and Northern Europe are likely to spread Mediterranean species. To prove this, an experiment was carried out some years ago where a car, whose tyres had been especially scrubbed clean and free from seeds, was driven 105km along ordinary roads after heavy rain, including a couple of stops in field gateways. It was then hosed down and the seeds collected and planted in sterilized compost. Thirteen species of flowering plant soon germinated, including 387 seedlings of annual meadow grass, 274 seedlings of chickweed and 220 seedlings of pineapple mayweed.

Oxford ragwort is a highly successful opportunist plant in towns, cities, railway sidings and around derelict buildings. It finds a living pushing out of cracks in pavements and brickwork and thrives in the urban environment where it is warmer than in the surrounding countryside. It can be told apart from the common ragwort by its yellow flowers which have black marks on the back of the florets.

Trains also play their part in seed distribution, perhaps the most famous example of all being the spread of Oxford ragwort throughout Britain. This plant was introduced from Mount Etna, Sicily, to the Oxford Botanic Gardens in the 1790s, its arrival coinciding with the rapid expansion of the railway system. Being used to the warm, stony slopes of its native volcanic habitat, the seeds escaping from the gardens soon prospered on the artificial stony wastes laid by man – the aggregates laid down for the railway sleepers. The passing rush of the trains undoubtedly helped to spread it further up and down the track, and it is thought that some seeds were wafted into the carriages only to be blown out again miles further down the line.

RAILWAYS

Europe's railway system provides a wide range of habitats for wildlife – a real linear nature reserve. With 18,000 km of track in active service in Britain, 28,000 km in West Germany and over 34,000 km in France, there is certainly plenty of room for colonization. Many defunct lines are richer in wildlife than active lines. Some have been turned into nature reserves, others designated as parks, but more have been reabsorbed back into the countryside so that no trace of them exists today.

Travelling by train provides a comfortable way to watch wildlife. Woods thick with bluebells and anemones, heathy areas abloom with the yellow of gorse and broom, and small riverlets banked by bulrushes (greater reed mace) flash by in constant succession. Many of the birds and mammals you might see from the carriage window, like magpies, rabbits and foxes, are so used to the rush and clatter of the trains that they carry on with their day-to-day business regardless of the locomotives passing so close by. The carriage window acts rather like a visual natural history programme, showing an ever-changing scenario. Different types of habitat flit by as the train crosses from one soil type to another, and different creatures present themselves for a

fleeting instant to the passer-by.

Apart from different soils and vegetation, the age of the track has a strong bearing on the kind of wildlife liable to be found there. Newly-dug embankments will sport a variety of grasses rather than the scrub of well-established lines and, as many railways embankments are over 100 years old, some are mature and thick enough to support a rich variety of species.

Birds are easily studied from a carriage window – and not only those that are seen near the train lines themselves. Passengers are often afforded clear views of those in the surrounding countryside as well. Rooks, crows, pigeons, starlings and lapwings can all be seen as flocks in

Above: *Wood anemones can blanket a woodland floor in early spring, their star-like flowers swaying in the wind, hence their alternative name of windflower. They can be told apart from other white flowers of the same habitat (eg wood sorrel) by their finely subdivided, three-part leaves*

Above: *Spring surprises await the gazer from the carriage window – perhaps a woodland floor carpeted with bluebells, or white with wood anemones and wild garlic. Cutting as they do, through a variety of natural habitats, railways often give the traveller the chance to look deep into the heart of the countryside – without the need to don boots or get wet!*

Left: *As soon as a railway station is abandoned many forms of wildlife move in. It does not take them long to colonize the steps, bridges, tracks, gravel dumps and platforms – all virgin habitats waiting to be exploited. Shrubs like privet and elder soon invade the stonework so that after ten years the undergrowth of bramble and scrub is so strong that it would appear to have been untouched by man or machine for half a century.*

nearby fields, working the turf or soil for small invertebrates. Seagulls can also be seen, often forming a prominent patch of white amongst the green or brown of the arable land. Near the track you may see magpies moving from tree to tree – very obvious with their black, white and steely blue colours. Last century, the naturalist Francis Buckland recalled how he saw "all along the railway from Boulogne very nearly to Paris...a magpie's nest in almost every tree", proof that the magpie is indeed an opportunist bird, quickly moving into a man-made habitat which had only come into being a few decades before.

Thickets of scrub provide plenty of breeding and feeding places for birds such as dunnocks, wrens, robins and linnets, and songposts for yellowhammers and blackcaps amongst others. The dense vegetation which has sprung up along the edges of many embankments also provides an ideal refuge for all kinds of bird, such as warblers, thrushes, tits and woodpigeons – as well as various mammals. Commuters are often amazed to see the sheer bravado of foxes walking

WHOSE NEST?

Magpie. *These birds have a great association with man, often building close to houses and factories as well as along railway lines. They are not called thieving magpies for nothing since they embellish their domed nests with bright objects. The nest is made of thorny twigs lined with turf and hair.*

Carrion crow. *Crows make bulky nests of twigs, mud and hair in tree forks. They will often use old nests, new ones taking some ten days or so to build.*

Woodpigeon. *These birds make a thin platform of twigs to hold their two eggs, often in the fork of an ivy-clad tree.*

Red squirrel. *Squirrels build dreys high up in trees in which to overwinter and rear their young. They are usually football-sized and built of twigs lined with grass. Dreys can easily be confused with bird's nests.*

Travelling by train during the winter gives you the opportunity to study nests in the hedgerows and trees. They are easily visible at this time of year, no longer concealed by thick vegetation. You may see nests where you least expect them, indicating plucky birds that will take advantage of nesting sites close to the roar and clatter of locomotives.

When trying to identify a nest, take note of whether it is single or part of a group, its shape and, where possible, the materials used to make it. All are aids to identification.

quite unconcernedly along the edge of the track or gambolling with their cubs on the railway bank. Their earths are often situated nearby, despite the vibrations caused by passing trains. Rabbits are another familiar sight along the track edge, providing ready food for the wily fox.

Badgers are not so often seen, although their regular tracks may bring them across the lines at night. Because of their habit of sticking to these ancient paths, badgers are particularly susceptible to changes in their environment. The electrification of railway lines earlier this century was no exception – scores of these amiable mammals were killed all over Europe. The same carnage can be seen on any new motorway that happens to cross their well-trodden paths.

If you look carefully you will see butterflies along the embankments of main-line railways. They tend to be the same species that you would expect to see along a country hedgerow, in an urban park or an old garden. Bramble blossom, a common sight at the edge of many tracks, does much to attract commas, small tortoiseshells, red admirals, common blues and gatekeepers. The

Above: *Much flora can be discovered by studying walls, like this one along the edge of an abandoned railway track. The brick platforms make an ideal substrate for colonization by plants and lichens. Here the widespread orange lichen* Xanthoria *is surviving on the mortar found between the bricks, and the fern wall rue has a firm hold along one of the cracks, its roots spreading inwards to the soil beyond. Ivy can also be seen, using the substrate to continue its relentless vertical climb.*

latter species (also known as the hedge brown) is particularly common in the summer. If you are passing over chalky soil, look out for the marbled white with its black spotted wings. Its caterpillars feed on scabious and knapweed.

Another obvious butterfly to look out for is the brimstone. Flying up and down the embankment, the male is very distinctive with his rich yellow wings. The female is harder to spot and, with her delicate lime green colouring, may easily be mistaken for a small or green veined white. Male orange tip butterflies are also commonly seen from the carriage window in the spring. Only the male has colourful splashes of orange on the tips of his fore wings – the female could easily be confused with many of the other white butterflies unless inspected closely. Both sexes, however, have a remarkable mottled green underside to their hind wings which helps to camouflage them perfectly amongst the vegetation.

Railway embankments often harbour an enormously rich array of wild flowers, the spectacle of which can be seen from the carriage window with the changing of the seasons. A botanist's delight can be found in the spring when yellow patches of primrose mix with the stitchworts, bluebells, lady's smock and wood anemones after the delicate green of dog's mercury has covered the mantle of the woodlands. These are soon replaced by knapweeds, hogweeds, escaped Michaelmas daisies, foxgloves and mulleins in the summer. Yellow patches of evening primrose, Oxford ragwort and St John's wort vie for attention with the red and purple of valerian, hemp agrimony, creeping thistles, fireweeds and the pink of everlasting pea. Riotous colour continues through the autumn with the colourful fruits of hips, hawthorn and ivy.

It is surprising what manages to burgeon out of seemingly impoverished stony ground near the edge of the track. Plantains and other low growing plants scrape a tenuous existence, and near city centres it is possible to see examples of the tree of heaven (a self-sown immigrant from China), colourful snapdragons, bushes of

Right: *Only the male of the common blue butterfly, seen here, is pure blue, a structural colour used for sexual recognition. The female is always brown or occasionally mottled blue. The males often congregate at 'watering holes' like moist banks or on mammal droppings to drink vital nutrients, however unsavoury.*

Far right: *The small heath can be exceedingly common along railway edges in the autumn. Like many of its relatives in the brown family of butterflies, it lays its eggs on grasses, and may be found in the same localities as meadow browns and ringlet butterflies.*

THE WILDLIFE OF A DISUSED RAILWAY

Railway lines epitomise the concept of linear nature reserves. They are green avenues for wildlife, abounding in wild flowers, insects, birds and mammals. Abandoned lines are soon engulfed by plants – invasive grasses that provide prime egg-laying sites for butterflies such as skippers and browns, and opportunist wild flowers that provide further food or hiding places for small flies and beetles.

Trees once cut back to ensure a clear track, now spread their branches ever upward and outward, harbouring nesting sites for woodpeckers, nuthatches and other birds, whilst foxes may make their dens amongst the undergrowth.

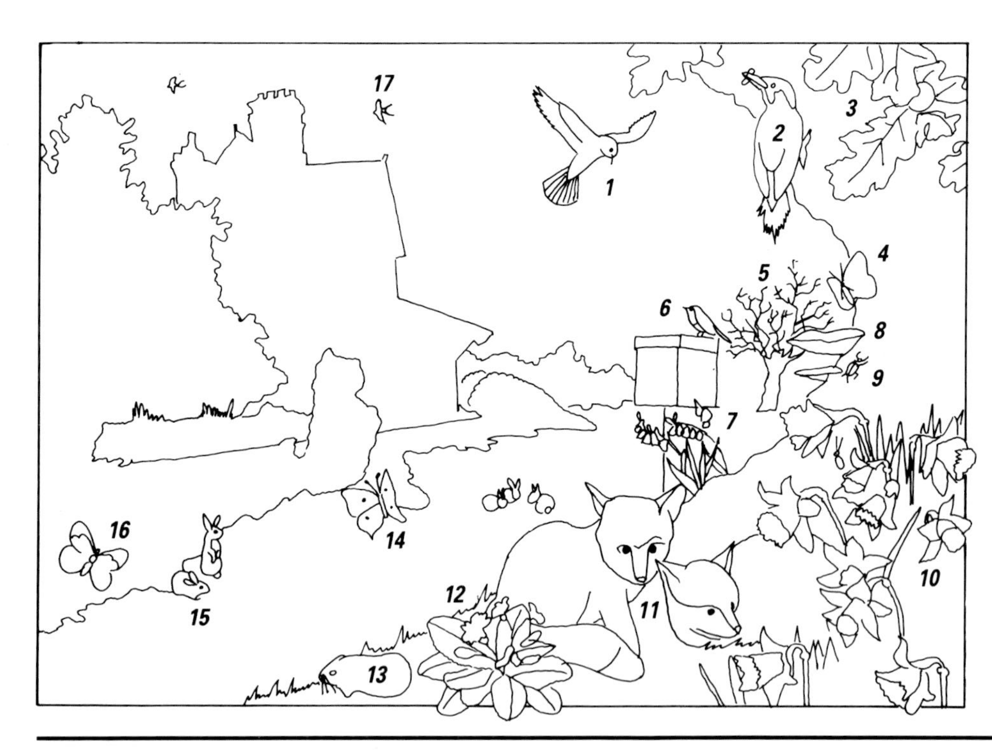

1 *Kestrel (male)*
2 *Green woodpecker*
3 *Common oak*
4 *Green-veined white butterfly*
5 *Apple*
6 *Robin*
7 *Bluebell*
8 *Beefsteak fungus*
9 *Stag beetle*
10 *Daffodil*
11 *Fox cub*
12 *Primrose*
13 *Field vole*
14 *Brimstone butterfly (male)*
15 *Rabbit*
16 *Orange tip butterfly (male)*
17 *Swallow*

Left: *Rosebay willowherb is now a familiar plant along roadsides, embankments and many other urban habitats. Each plant can produce as many as 80,000 fluffy white seeds which are carried by the wind to new ground. This plant is also known by its American name of fireweed after its remarkable ability to colonize freshly burnt ground. Its leaves are often eaten by the impressive caterpillars of the elephant hawk-moth.*

buddleia alive with butterflies and tall stands of the introduced Japanese knotgrass.

Different plants and shrubs along the edge of the track signal the different types of soil. In chalky areas the embankment can be covered with old man's beard, yew, spindle, whitebeam, wild roses and wayfaring trees. On sticky clay, oaks, sycamore and ash flourish – so much so that they sometimes knock out power lines or get in the way of passing locomotives. In sandy areas look out for gorse, broom, bracken and the attractive silver birch.

In some areas the relics of railwaymen's gardens still adorn the embankments. Wild apple and pear trees, currant bushes and raspberries run wild, together with roses, poppies, lupins and all manner of other cottage garden plants.

ROADS

Everyone in Europe can see wildlife along major trunk roads, motorways, autoroutes and autobahns. Look for the kestrels hovering over the carriageways, the bright seasonal splashes of wild flowers on the banks and the scuttling of a weasel or field mouse amongst the grasses of the verge. Wildlife has the uncanny ability to colonize all kinds of alien environments, and road and motorway edges are not nearly as alien as many other man-made places. They represent long slithers of natural grassland which may become the last stand of species once found commonly throughout the countryside.

Ancient roads are often bordered by hedgerows which have been in existence for many decades. By counting the number of woody plant species growing in a 27-metre stretch of hedgerow you can arrive at a fair approximation of its age – each species representing a hundred years.

New roads, particularly motorways, invariably involve the creation of banks and verges of freshly turned earth – sites soon colonized by a multitude of grasses and wild flowers. Despite various problems – pollution, danger from passing traffic, and in some cases incessant noise and vibration – wildlife continues to flourish in these linear reserves. Not least, the birds...

Above: *Much of the richness of a roadside walk was lost with the advent of tarmac and then, more recently, verge and hedge cutting and the use of herbicides to control the spread of weeds. However, lanes and tracks in rural areas are often still rich in flora and a walk or cycle ride along such will certainly have its rewards and be more productive than travelling by car.*

Below: *Newly-prepared motorway embankments are soon exploited by opportunist plants. When the soil is disturbed, seeds from local crops such as oilseed rape are blown in and germinate freely. Other wild plants such as challock or poppy may spring up from seeds blown along the carriageway from other areas or from seeds buried deep in the soil being shifted to the surface. The resulting effect can sometimes be quite unexpected and often startlingly colourful.*

THREE BLACK BIRDS

Rook. *The rook can be distinguished from the crow by its pale grey beak with a bare white patch at the base. It also has a more purplish sheen to its head feathers.*

Crows. *Crows are big birds entirely covered with glossy black feathers. They have heavy bills which are also black.*

Jackdaw. *Jackdaws are smaller than rooks or crows and can be distinguished by their grey nape.*

Several black birds are commonly seen in groups and can be hard to distinguish. Starlings, rooks, crows and jackdaws, even choughs may be seen, but how do you tell one from the other?

Starlings (and blackbirds for that matter) are the smallest of these black birds and can be recognised by their yellow-orange bills. In winter, the starling's plumage develops a white speckled breast. Choughs are easily identified by their bright red legs and bill. They are also much less common than the other species, being found on sea cliffs and rocky islands, and on some urban rubbish dumps.

ROADSIDE BIRDS

Travellers along major roads, particularly motorways, cannot fail to miss the hovering kestrels, the flocks of birds in adjacent fields, the irregular swooping flight of the lapwing and the rooks, crows and magpies that scavenge at the side of the road. These are just the obvious ones – many more remain concealed in the undergrowth beside the verge, or keep a look out from their perches on nearby posts.

Roads provide a hunting ground for many different species. In Southern Europe buzzards and red kites can be seen flying up and down the motorway verges in their search for small mammals. Kestrels can be seen too, the only hovering birds of prey, checking out the central reservation or verge in their search for some unwitting victim. So used to man have these birds become that they can sometimes be seen sitting on motorway signs, bridges or roadside posts.

Smaller birds have adopted the grassy verges as ideal nesting sites. Corn buntings, meadow pipits, pied wagtails and skylarks breed amongst the grasses, whilst blackbirds, thrushes and robins frequent the bushes and other undergrowth. The skylark in particular can often be seen hovering close to the carriageway, waiting to descend but anxious to keep its nest site a secret. Ground nesting birds such as pheasant and partridge may also nest in the long grass which grows up in the early summer.

ROADSIDE BIRDS	
Blackbird	Kestrel
Black-headed gull	Lapwing
Buzzard	Magpie
Chaffinch	Pied wagtail
Carrion crow	Red kite
Feral pigeon	Rook
Hedge sparrow	Robin
House sparrow	Sand martin
Hooded crow	Skylark
Jackdaw	Starling

Service stations are another good place to study birds. Chaffinches, pied wagtails, house sparrows and starlings rush forward to take the crumbs and waste food left by travellers before the rooks, crows and magpies descend to claim their share of the pickings. These bigger birds can also be seen pulling and tugging at rubbish bags in the hope that they will reveal some tasty morsel. Indeed, no highway would be complete without the magpie – an opportunist scavenger that will often alight in the middle of the road to pick at the corpses of small birds, rabbits, hedgehogs and squirrels.

Right: *Many birds nest amongst the grasses and undergrowth of wide roadside verges. Walkers may accidentally come across the well-camouflaged nest of the red-legged partridge, a native of south-west Europe. This bird will wait until the last moment before deserting its nest, bursting forth from the vegetation with a wild flurry of wings. Once disturbed, it is said never to return to its nest and eggs.*

Below: *Resourceful rooks often have to compete for food with magpies, crows, sparrows, pied wagtails and starlings – all birds of motorway service stations and roadsides. These birds have been known to store food under grass cuttings when a surplus is available and can frequently be seen with their beaks stuffed full of scraps scavenged from litter bins.*

Crows and rooks have also learnt to take advantage of another food supply, by developing an ingenious way of prising squashed food off tarmac! Any food that is dropped or thrown out of vehicles is usually run over and squashed fairly effectively into the ground. The same applies to the corpses of animal and bird victims. With head and bill sideways to the carriageway, these birds employ a scissor-like approach to prise the food off the ground. In doing this, the birds have also developed a considerable amount of road sense – presumably evolved through natural selection of the fittest and brightest individuals. On entrances to service stations rooks often alight facing the traffic, for example. Their sheer nerve in the face of high-speed vehicles is quite amazing – particularly on motorways. It is a common enough sight to see birds scavenging food only feet away from the deathly tread of 130kph tyres. There are the inevitable casualties though. These are the ones who looked the wrong way or moved at the wrong moment, for there is a distinct advantage in staying still if you are just outside the bounds of the white lines. The evolution of such nerve-wracking behaviour only comes about by natural selection and goes to prove that wildlife has a great ability to make adaptations to the curious habitats provided for it by man.

MAMMALS

By creating grassy embankments and verges along the edges of many carriageways man has created the preferred habitat of many small mammals. Ordinarily they are found in orchards, meadows and other places where there is plenty of grass to conceal them from predators, but the roadside verge offers them another alternative – long thin grasslands that seem to go on for ever. In many cases, living inside the confines of the roadside is the only alternative. All

WEASELS AND THEIR RELATIVES

Weasel. *These ferocious carnivores are smaller than stoats with a more irregular line between their chestnut upper parts and white underside. They are most often seen dashing across the road but two weasels fighting amongst themselves may sometimes be heard – a very noisy affair indeed.*

Stoat. *Stoats can be easily identified by the black tip to their tails. They are bigger than weasels and in severe winters turn white to give the appearance of ermine. The black tip to their tail is never lost.*

Pine marten. *These creatures are bigger than stoats and weasels, being about twice the size of a squirrel. Their rich dark brown fur gives way to a paler yellowish bib, and their bushy tails and large feet are great aids to climbing. They are rarely seen, being nocturnal and very wary of man.*

Nearly half of Europe's carnivorous mammals belong to the weasel family or *Mustelidae*, including the polecat, badger, mink, otter and glutton, as well as the stoat, weasel and martens.

Stoats and weasels are widely distributed throughout Western Europe and may occasionally be seen during the day – frantically dashing across the road to hunt small mammals amongst the grasses of the verge. Pine martens are more usually seen in forested areas on their nocturnal wanderings, picked up in the headlights of the car. Unlike their relatives, the martens have perfected the art of climbing, and are agile enough to chase and catch squirrels amongst the branches.

that awaits them in the surrounding land is a hostile environment of houses, factories and intensive agriculture, so it comes as no shock to find that many mammals have been quick to exploit these new grassy habitats.

Harvest mice have been found living on motorway verges within one year of their opening and other small mammals, including mice, shrews and voles, live in tunnels amongst the long grass during the summer. We seldom see these creatures unless they make a frantic dash across the road, but the numerous kestrels hovering above the verge testify that plenty of them must be there.

It is unfortunate that a high proportion of the mammals we see on roads and roadsides are dead ones – something which goes hand in hand with the fact that so many creatures use these green corridors. Some make them their permanent homes; others traverse them in an attempt to simply cross the road. Badger tracks, for example, existed thousands of years before those of man and today these harmless creatures are being killed in the same old places where generations of them have plodded before. We may disect their territories countless times with roads and railways but they will still endeavour to cross them, rather than changing their age-old habits.

Newly-opened motorways always create a carnage of squashed badgers and foxes. This is partly because the animals get used to crossing the carriageways before the motorway is opened, so when the vehicles arrive they are taken unawares and many are killed. In some parts of Europe deer, wild boar, pine and beech martens and wild cats are an additional hazard. The wild

Above: *The sika deer is a native of Eastern Asia but is now found in many parts of Western Europe. It is larger than the muntjac which also comes from the east, and can be distinguished by its black and white rump. This juvenile buck has yet to develop the large, many-pronged antlers of the adult.*

Sika deer normally live in small herds, emerging from woods during darkness to feed in grassy areas. They sometimes stray onto adjacent roads and may occasionally be seen on the verge.

Above: Self-explanatory notices along many of the autoroutes of France and other European countries detail some of the larger mammals and birds that may be seen whilst driving. Other forms of wildlife depicted on large signs include rabbits, squirrels and birds of prey.

Right: The wild boar is widespread in mainland Europe but is a very wary traveller. You may be lucky enough to see it crossing the road at night. An ancestor of the farmyard pig, these creatures live in wooded areas where they grub around for roots and bulbs. The old boars may be solitary and have fearful tusks which they use for defence. The sows are just as fierce when defending their young.

boar is widespread in mainland Europe but is a very wary creature seldom seen by man. However, in rural areas boars, like badgers, have definite, well-worn paths which sometimes cross roads at specific points. Weighing up to 340 kg, collisions with motor vehicles can have serious consequences for both parties.

Many attempts have been made at finding a solution to the problem. One successful operation has been to instal tunnels for badgers and wild boars underneath those motorways that are directly in the path of their well-worn tracks. In forestry areas special roadside mirrors have been erected which deflect the lights of cars into the surrounding woods to deter deer from coming too close. Deer fences, 2 metres high, have also been erected to some effect, but in some areas the animals have become fatally impaled.

Watch out for small mammals – hedgehogs, rabbits, weasels and various rodents – running across the road in front of the car at night. In some cases it will be the only chance you'll get to see them.

ROADSIDE MAMMALS	
Badger	Harvest mouse
Bank vole	Hedgehog
Common shrew	Rabbit
Field vole	Wood mouse
Fox	Yellow-necked mouse

FLOWERY VERGES

Many attempts have been made to brighten up the verges of our motorways and roads by planting various wild flowers along them. Aquilegia (columbine), autumn and spring crocus, snake's head fritillary, great bellflower and Jacob's ladder, wild daffodil, wild tulip, foxglove, primroses and cowslips are just a few of the species that have been introduced to selected motorway verges in Northern Europe. Even more become packed with brightly-coloured flowers without man's intervention.

Great patches of pink announce the presence of early purple orchids, common spotted orchids and pyramidal orchids in limestone areas, forests of umbellifers (hogweeds, hemlock and cow parsley for example) stand in vast thickets down the sides of many roads in the spring and summer, and brambles sprawl endlessly down the hedgerows. In fact, it is thought that at least a third of our wild plants are now happily growing along roadsides. Most have got there by themselves, including some rarities – spiked rampion, purging flax, yellow vetchling, dark mullein and pepper saxifrage to name but a few. But it is the grasses which form the predominant vegetation along many verges. You might find over 50 different species, most of which will have colonized the roadside by themselves.

Other colonizers are governed by both the age of the verge and the type of soil. On light

ROADSIDE WILD FLOWERS	
Barren strawberry	Lizard orchid
Bee orchid	Melilots
Bird's foot trefoil	Old man's beard
Bitter vetch	Plantain
Coltsfoot	Pyramidal orchid
Common spotted orchid	Ragwort
Dandelion	Toadflax
Early purple orchid	White butterbur
Fennel	Winter heliotrope
Fly orchid	Yarrow

sandy soils, white clover soon invades the grass to create patches of darker green and provides a feast for honey-bees in July. After some time the mass of clover soon succeeds into a vegetation dominated by broom and gorse. After five years the once-neat grassland becomes fragmented into areas encroached by bramble scrub, broom, gorse and semi-mature trees which, in some cases, have been planted by man.

One of the characteristic plants along ordinary verges is coltsfoot. Its bright yellow flowers can first be seen in February and March, sometimes pushing their way through snow to reach the light. Coltsfoot is extraordinary in that its flowers and leaves appear at completely different times – the flowers emerging in the spring, followed by the untidy leaves in summer and autumn.

Another common roadside plant is winter heliotrope. This is a Western Mediterranean species which has the ability to colonize roadside verges very effectively, often to the exclusion of all other species. Its large leaves survive throughout spring, summer and autumn and will grow vigorously if cut. This heliotrope, meaning sun-worshipper, also flowers early in the

Below: *A wide variety of plants can be found growing along the roadside or in waste areas. Here are some interesting ones. The henbane (below left) has poisonous fruits, the active alkaloids of which are used in a variety of medicinal drugs. They act on the nervous system and are employed in air and sea sickness tablets.*

Hops (centre) belong to the cannabis family and are native to much of Europe. Extensively employed by the brewing industry, they are grown in both Germany and England and have escaped into many hedgerows where their perennial roots push up vigorous shoots every year.

As a symbol of the Roman Empire, the leaves of bear's breeches (below right) can be seen carved into many stone columns of the period. The plant still survives around old ruins, castles and wasteland as well as being a familiar sight in gardens.

Left: *Trees like the London plane are often used to line roads in mainland Europe, providing shade and also acting as windbreaks. In some city centres these and similar forest trees can cause considerable problems, undermining pavements, walls and buildings.*

Right: *The giant hogweed is a distinctive plant growing to heights of over 3 metres. It is native to Russia but can be found on roadside verges and central reservations throughout Europe. The sap which oozes from its hollow stem is a skin irritant.*

year, sometimes in the snow, and for this reason was introduced to Northern Europe to brighten up otherwise dreary gardens during the winter. The flowers are a delicate pink and are very fragrant. Around the Mediterranean the true heliotrope can be found. This straggling plant with its groups of white flowers, occurs on recently-turned soil at the wayside or on dumps of gravel or building materials.

Roadside verges are often ablaze with colour in the spring. Bluebell, greater stitchwort and red campion often grow together to give a red, white and blue effect. Many plants occur like this as an association of certain species growing together. This is one of the most common.

Other roadside plants to look out for include the brightly coloured bitter vetch which straggles up through the long grass and climbs the hedgerow. Similarly, red byrony, old man's beard (also known as traveller's joy or wild clematis) and honeysuckle are all perennial hedgerow climbers that give marvellous displays of colour in the autumn. Speedwells, willowherbs, campions, comfrey and chicory are complemented by the mass of yellow flowers found along the verge. The ragworts, groundsels, hawkbits and dandelions all share a golden coloration and their fluffy seeds are perpetually blown along the roadside by passing traffic.

Below: *Orchids are common beside roads. It may be that verge-cutting actually helps these flowers to prosper, since long grasses soon swamp these delicate species. Moreover, if their flowering spikes are inadvertently cut, it may stimulate further growth the following year. Pyramidal orchids, shown here, can often be seen as patches of pink blooms during spring and summer.*

THE SEASIDE PLANT INVASION

Perhaps one of the most extraordinary facts to emerge about roadside vegetation has only come to light in the past decade. In the 1970s it was noticed that some wild plants which are normally found only on coastal salt marshes were growing along the banks of motorways and major roads in Britain, East and West Germany, Northern France, the Netherlands and North

America. What appears to have happened is quite simple – man has succeeded in creating a novel artificial habitat thanks to his use of de-icing salt. So much of this is distributed on carriageways during freezing weather that it has altered the soil chemistry of the roadside verge. Instead of disappearing between winters, the salt accumulates, with the result that there are now verges which are 50 times more salty than coastal grasslands normally subjected to salt-spray.

Only certain plants have tolerant properties that allow them to thrive in salty conditions and three such species have surreptiously taken to the roads from their seaside retreats – a group of grasses called the sea poas, sea plantain and sea spurrey. But how have they succeeded in travelling so far inland to colonize these new habitats? Again, they have been given a helping hand by man – being carried in the mud stuck in tyre treads, under wings and behind bumpers only to be deposited in a similar environment many miles inland! The colourful sea aster has also made the journey and can occasionally be seen on motorway embankments.

Some species of grass appear to be more tolerant of high concentrations of de-icing salt than others. Red fescue, rye grass and crested dog's tail seem to have made better adaptations than white clover and meadow grass, and there is some indication that other common roadside plants such as mugwort, broad-leaved dock and knotgrasses have also developed a degree of tolerance to salt. Brown patches of vegetation – a tell-tale sign of burning by salt – are still a common sight however, often creating areas where only hardy plants like the sea poa can survive.

Right: *Ivy provides good cover for insects and other invertebrates, as well as rich helpings of pollen for overwintering honey-bees. It is a tenacious climber which will smother trees, walls and stonework alike. Letter-boxes provide an equally suitable habitat, thereby providing a cave-like atmosphere for all manner of creatures, including snails, slugs, woodlice and earwigs. Snails can prove a problem in this respect since they will readily feast on paper envelopes in the same way as they do on leaves.*

Left: *The buck's horn plantain is normally a coastal species found throughout most of Europe, but has ventured inland following the salt-rich verges that result from the frequent use of de-icing salt. Its name is derived from the antler-like rosette of leaves at its base.*

Below: *The bright colours of ladybirds – either black and red or yellow and black, according to the species – act as a deterrent to predators, indicating the fact that they contain poisonous chemicals. They are common on many wayside plants where they feast on aphids. In the winter they hibernate in available cracks and crannies in wooden posts and under rocks.*

INSECTS ALONG THE VERGE

Insect populations along the verges of ordinary roads and motorways can be exceedingly high. Ladybirds are a frequent sight, resting on the broad feeding platforms of hogweed, concealed on the undersides of flowers or searching out the aphids that cover so many roadside plants each summer. Ladybird larvae can also be seen – long brownish creatures not often recognised as an earlier stage of these pretty beetles – and they are just as predatory as their parents. Look for other types of ladybird, apart from the usual red and black 2-spot. Some are yellow and brown or

black, and others have as many as 24 spots on their tough wing covers.

Grasshoppers and crickets usually abound in the roadside grasses. The dark bush cricket is a common species which is easy to find thanks to its distinctive chirping sound, usually made in early evening. Walk through the roadside grasses in Southern Europe and you will disturb plenty of grasshoppers that will expose the blue or red flashes of their hind wings as they fly off. The great green bush cricket can also startle the unwitting passer-by. Being green, it is normally well camouflaged but, when disturbed, it will

ROADSIDE BUTTERFLIES	
Common blue	Meadow brown
Dingy skipper	Peacock
Gatekeeper (hedge brown)	Ringlet
Green hairstreak	Scarce copper
Green veined white	Scarce swallowtail
Grizzled skipper	Small copper
Large skipper	Small skipper
Large white	Small tortoiseshell
Mallow skipper	Small white
Marbled white	Wood white

A GUIDE TO SOME HEDGEROW INSECTS

Insects are the most numerous animals along the hedgerow and verge. They bask on leaves, forage for pollen or nectar, and chase or kill other insects for food. Some familiar roadside insects are shown here. They may also turn up in the herbaceous borders of parks and gardens.

Polistes wasp. *A slimmer version of the common wasp which is confined to Southern Europe. It builds a small nest which hangs down from vegetation or rocks and may also be attached to window arches and shutters.*

Hoverfly. *Hoverflies are very numerous in spring and summer. Some are coloured like honey-bees, others like wasps. This is all part of an elaborate mimicry designed to protect them from predators. In fact they are quite palatable, and have no sting.*

Click beetle. *The adult form of the wireworm (a serious agricultural pest), these beetles can propel themselves away from danger by suddenly reflexing their abdomen. This action also makes a loud click, hence their name.*

Greenbottle. *A true fly which has bright green livery, and is seen in a variety of habitats. It lays its eggs on dead animals, the larvae helping to degrade the organic matter. A useful and colourful insect to have about.*

Leaf hopper. *These insects belong to the same group as froghoppers, aphids and scale insects. They jump from leaf to leaf and many are dressed in exciting colours.*

Left: *Scarce copper butterflies are common on rural roadsides in much of Europe except Britain. Their brilliant colour – more obvious in the male – makes them easily identifiable, and they can even be seen investigating the bright orange plastic of car indicators believing them to be prospective mates.*

Below: *A female meadow brown butterfly feeds on the nectar of a carline thistle, displaying the false eyes meant to deter predators. Meadow browns are one of the most common butterflies in Europe, their caterpillars feeding on grasses which are always abundant on roadside verges and embankments.*

heave its 9cms frame into action and fly off for some distance before alighting in the vegetation. Field crickets can also be found lurking in their specially prepared holes in the ground. They have much shorter antennae than most crickets.

Other insects that you may come across along the roadside verge or hedgerow are the shield bugs. The most distinctive feature of these curious beetles is the possession of syringe-like

OTHER ROADSIDE INVERTEBRATES	
Assassin bug	Ladybirds
Brown tail moth	Oak eggar moth
Buff-tip moth	Preying mantis
Bumble bees	Puss moth
Cinnabar moth	Sawflies
Field cricket	Shield bugs
Freshfly	Soldier beetle
Grasshoppers	Vapourer moth
Great green bush cricket	Wasp beetle
Humming bird hawk moth	Yellow ant

mouthparts which are designed for the drinking of sap, quite unlike the biting jaws of ordinary beetles which work from side to side. There are many different species, some being named after their food plants (the gorse and birch shield bugs for example), and others which have a distinctive red and black coloration designed to ward off predators. The old fruiting heads of wild carrot which can be common along roadsides often support a considerable number of these insects. They are closely related to the aphids and cicadas which also have piercing mouthparts.

Watch out for the slim parasitic wasps with their long 'tails' (which are really ovipositors – egg-laying devices) which hunt among the vegetation for caterpillars which they paralyse with their sting and then use as live food in which to lay their eggs. The predatory robber flies can also be seen, chasing other flies to suck out their fluids. Dragonflies also course up and down the verges chasing insects, and waiting

Above: *The dark bush cricket is common along roadside verges as well as in the long grass of orchards and gardens. Large populations are often evident simply by the sound of their incessant chirping, particularly during the evening. Unlike grasshoppers, they prefer to walk through vegetation rather than jump.*

inside the flower heads are spiders and praying mantises, often deceptively camouflaged against their backgrounds.

Over and above birds and mammals, insects receive the highest mortality rate along roads. It has been estimated that about 1,800 million moths are killed each year after dark on British roads alone, and this represents only one order of insects. The toll taken in the rest of Europe, on the autostrada, autobahns and the French autoroute de soleil must be colossal. Many of them fall, dead or simply stunned, to the side of the road which is the beginning of an interesting sequence of events.

Most of the insect remains, dead or alive, are blown to the edge of the road where they become entangled in the grass of the verge, and this means big business for the ants. They, like so many other creatures, have learnt to cash in on this fast food supply delivered to their doorstep with every passing vehicle. In some places along motorways, you can find colonies of yellow ants living every few feet along the edge of the hard shoulder. This interface between road and verge is therefore a very important place. Birds, too, can be seen mopping up the insect casualties and what they miss, the ants gain!

THE MOTORWAY HABITAT

Motorways have been in existence for less than thirty years in Western Europe. They have been cut across mountains, hills, lowlands and wetland areas and the surrounding wildlife has increasingly infiltrated the man-made embankments and verges.

Travelling fast across country by motorway gives everyone the chance to study the interaction of wildlife from farms, woods, moors and fields with the new habitats provided by man. Certainly wildlife seems to thrive along these green corridors, but we must not forget that they have replaced natural habitats – meadows, brooks and heaths all destroyed in an irreversible earth-moving exercise. It takes about 8 ha (20 acres) of land to build 1 mile of motorway and only about 40% of this land is returned to green verges. Habitats with rich colonies of butterflies, wild flowers and locally

CRICKET OR GRASSHOPPER?

Most long grass will provide a habitat for some grasshopper or cricket during the summer. The sound of their stridulations – each a unique melody of chirps sufficient to identify a species – may be very loud. Crickets 'sing' by rubbing their front wings together, grasshoppers by rubbing their hind legs against their wings.

Both insects have wings, although in some species these are very small and cannot be used for flying. They also have very long back legs which are used for jumping, and antennae which are considerably longer in the crickets. Grasshoppers' antennae are only about one-third the length of their bodies.

Field cricket. *Found along paths and in meadows where it makes a small burrow. It is sometimes kept as a pet.*

Meadow grasshopper. *One of the commonest species, found in many different habitats. Large populations may occur in very small areas.*

Great green bush cricket. *One of the largest green crickets, this insect will fly off for some distance before landing when disturbed. Its large green wings and long legs are an impressive sight.*

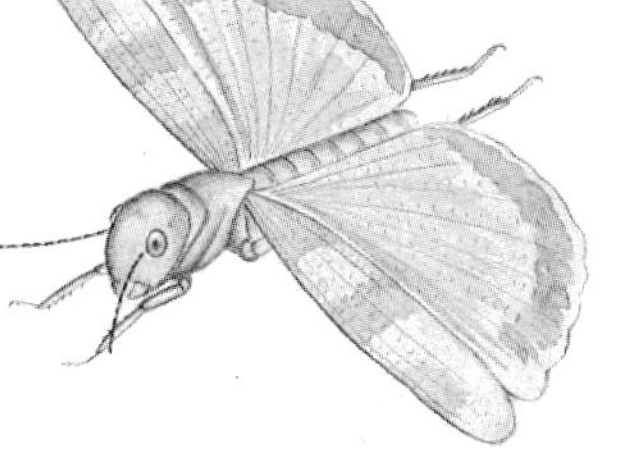

Oedipoda grasshopper. *These are found in Southern Europe. When they fly, they reveal their brightly-coloured hind wings – either red or blue.*

Left: *Many rare plants and invertebrates can be found living in the shadow of motorway flyovers. The spaces left to allow expansion and contraction between the prefabricated blocks prove particularly attractive to sparrows which can be found rearing their young amidst the clattering of overhead traffic.*

Below: *Fairy shrimps are common in roadside puddles as well as pools in coastal rocks. Despite adverse circumstances, they have learnt to survive in this alien environment, their eggs remaining unharmed in dried-up puddles waiting to resume development when the rains return. It takes the fairy shrimp only two weeks to reach maturity from the egg stage. They may be as large as 3cms.*

or nationally rare species disappear without trace – they cannot be spontaneously recreated overnight, even on aesthetic-looking motorway verges.

That said, there is still a surprising diversity of habitats on motorways which prove very attractive to wildlife. Some will not be noticed by the driver, but they can be roughly divided into two distinct areas – the central reservation and the grass verge or embankment. At the edge of the motorway there may be a boundary hedge which grows unmanaged over the verge. Older motorways have thickets of bramble, and great swirls of vegetation which provide nesting places for birds. Various amounts of scrub can also be seen on verges where plant succession has progressed. With regular tree-planting, vast patches of young woodland become established in a few years and oddly-shaped areas of 'no-man's-land' between motorways and slip roads often escape management and form wild sanctuaries for small birds, mammals, insects and plants.

Freshwater habitats occur too, where regular amounts of water run off into special holding pools. There are even a few such pools that are permanently filled with water and here aquatic insect colonies swiftly build up. Where streams flow under motorways it is quite possible to find 'motorway trout' – escapees from fish farms that live quite happily in the culverts provided for the water. You can never be quite sure whilst driving along of whether you are crossing a badger underpass or a mini trout reserve!

The great thing about motorway habitats is that they are allowed to prosper and mature without the influence of man, the picker, collector and trampler. Wildlife can really thrive in these areas because it is left alone. Today's motorway management is confined mainly to keeping sight lines open; that is cutting back any vegetation that overhangs the carriageway or obstructs the line of view between the driver and any sign. Occasionally the vegetation on the central reservation has to be cut back since it does so well, and large plants like melilot and mugwort tend to overhang the tarmac. Particularly colourful central reservations exist in Southern Europe where the pink, red and white varieties of oleander, tamarisk and mimosa mix with the viburnums, berberis and roses which provide bright autumn fruits.

POLLUTION AND OTHER PROBLEMS

There are four main pollutants along the edges of roads and motorways – mercury, lead, nickel and cadmium. Two others are dust and salt. All of these accumulate in or on vegetation, and in plant-eating insects and mammals but do not always harm them as one might expect. In fact, there is evidence to suggest that certain pollutants actually encourage plants and animals to grow!

The caterpillars of the buff-tip and gold tail moths can sometimes be found in large numbers stripping the leaves from roadside trees and bushes. They attack a wide range of species all of which are thought to absorb the nitrogen dioxide given off by passing vehicles. This nitrogen is one constituent of the amino acids that make up any form of protein, so it follows that the pollutant actually acts as a plant growth promoter, resulting in flourishing colonies of caterpillars.

Research work has also shown that dust from traffic settling on roadside vegetation increases the internal temperature of the leaves. This in turn increases their rate of metabolism and therefore the rate of growth and ageing. It is certainly true that roadside plants often grow and age quicker than their contemporaries in less polluted areas. This high turnover of plants has a consequent knock-on effect on insect colonies which have found the roadside a very suitable habitat indeed.

Lead, zinc and cadmium all accumulate in wild plants, mosses and liverworts, snails and small mammals. Lead certainly may then be passed on to any predators – kestrels feeding on verge-dwelling voles for instance.

The fact that such creatures can tolerate higher levels of toxic substances than they would experience in the countryside is yet another indication of their capacity to adapt to man's environment. Being able to make the most of new habitats and phenomena – whether by mopping up traffic casualties or colonizing the concrete wastes of intersections – ensures the continued success of many species along these green corridors.

PLANTS OF NO-MAN'S-LAND

Places left alone by man soon become colonized by wildlife, and the no-man's-land between motorway and slip road is no exception. Opportunist plants soon move in, thanks to their efficient methods of dispersal which are sometimes given a helping hand by passing vehicles – the fluffy seeds of plants like dandelions or willowherb are wafted along the roadside by the traffic or get caught up in the mud stuck in tyre treads.

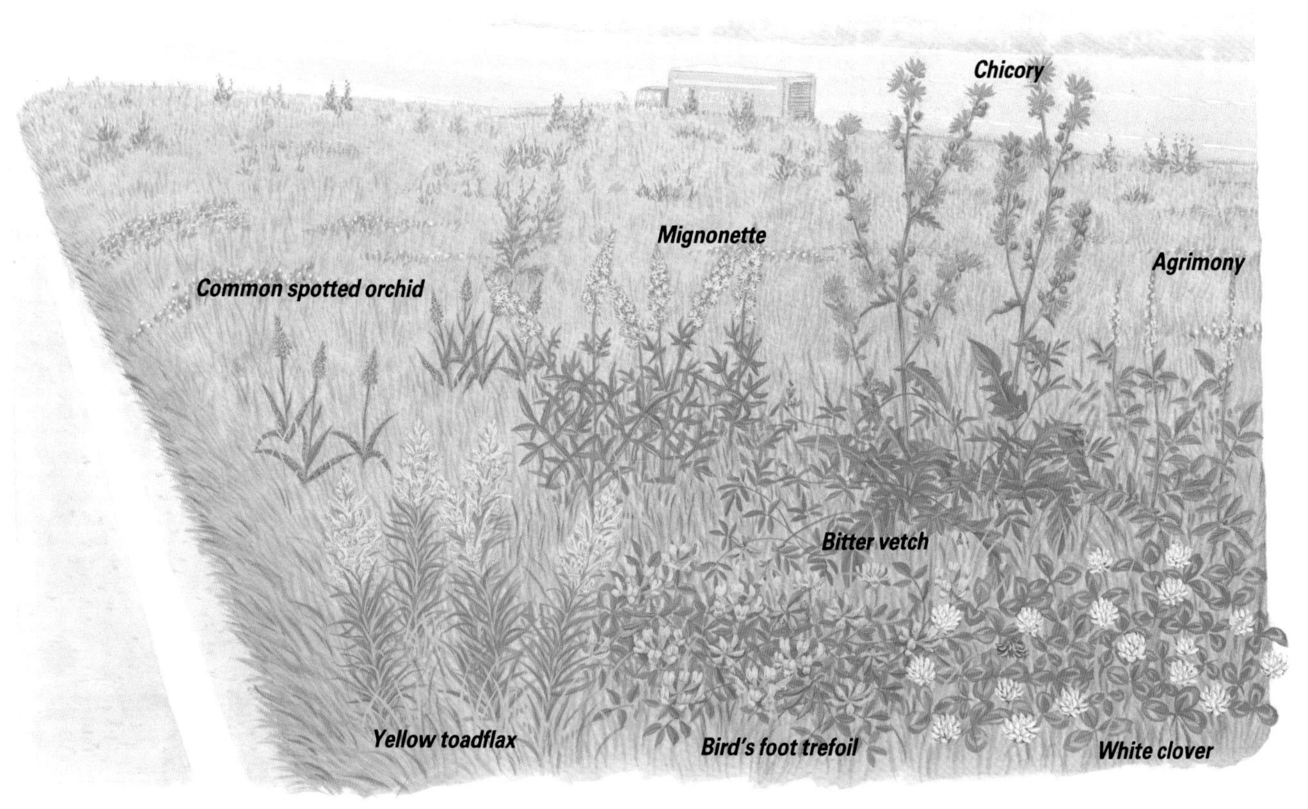

Canals frequently offer rich habitats for wildlife. The water is often crystal clear and, when examined, can reveal the delights of many small aquatic insects – whirligig beetles buzzing madly about in groups, diving beetles glittering from the air trapped around their hairy bodies and water boatmen darting amongst the weeds. The towpath itself may be knee-deep in wild flowers and there is always the chance of discovering an interesting species that may have been originally introduced to the habitat from the hay-bag of a towing horse.

CANALS & RESERVOIRS

Man has fashioned most of the bodies of open water throughout Europe. He has influenced the land with his gravel pits, reservoirs, ponds, lakes and canal systems, and in doing so has created yet further man-made habitats for wildlife to colonize. Introduced species grow side-by-side with native flora and fauna around the edge of lakes which were once the scene of great industry and along canal banks no longer washed by the passing of port-bound tugs. Today, rushes conceal the long-abandoned jetties, reed mace sprouts from the holds of old gravel boats and migrating wildfowl nest on artificial islands built by man in the middle of equally artificial lakes. Here, as in many newly-created habitats, wildlife continues to survive, thrive even, with the result that these watery domains support a teeming variety of life – life often found nowhere else but on your doorstep.

CANALS

Canals are linear nature reserves which thread across the countryside, linking distant seas, tunnelling through uplands and dissecting all manner of different habitats. As man-made transport systems they have acted, like roads and railways, as convenient corridors along which plants and animals can disperse and colonize. Quite apart from the waters of the canal itself, all sorts of other mini-habitats can be found – much of the machinery of canals, the walls, lock gates, towpaths, bridges, feeder reservoirs, cuttings and embankments serve to provide a varied sanctuary for flora and fauna alike. The result – waters and banks that now burgeon with wildlife, and to see it, all you need to do is take a relaxing cruise along your local waterway or a leisurely stroll along an old towpath.

Canals provide an attractive home for many reasons. Aquatic plants and water weeds, for example, grow so well in canals that they need to

Left: *Arrowhead is an exceedingly abundant emergent plant which is found in canals, ponds and lakes. It has attractive leaves and delicate spikes which hold the many white flowers. Often a mass of arrowhead will form at the water's edge, providing thick cover for water birds – like coots and moorhens – as well as a resting site for insects.*

Below: *Gipsywort can often be found growing from canal banks and waterside walls. The loose mortar between the bricks provides sufficient space for its seeds to germinate and plants to prosper. This plant was said to have been used by European gipsies to stain their skin black, hence its common name.*

be cleared frequently to prevent obstruction. The slow movement of canal water provides them with a better environment in which to grow than the quicker moving waters of most rivers. It is not surprising, therefore, that disused canals soon become engulfed by weeds and in doing so provide rich wildlife habitats for many kinds of invertebrate. This in turn attracts numerous species of fish, bird and mammal. With the demise of the canal system last century, there has been much opportunity to create canal nature reserves.

Apart from providing a refuge in urban areas, canals are also a valuable means of dispersal for both animals and plants. Some have used the boats themselves as platforms to aid quick dispersal while others have employed their own powers to spread along these artificial waterways. The spread of some plants along canals can be compared to the spread of their relatives along ordinary rivers and streams, except that with canal transport the process of dispersal is increased. The wash of boats along the canal banks increases the chance of seeds being washed into the water from where they will be carried downstream. Moreover, those creatures or seeds which inadvertently use the boats to aid dispersal can colonize new pastures up or down stream according to the direction of their transport. For them, the whole canal system

is there to be exploited.

Extensive canal systems were built throughout Europe during the eighteenth and nineteenth centuries to overcome bad road conditions, and although they were soon to become outdated by the railways, their presence across the landscape has remained to provide a quick throughroute for plants and animals between countries such as Belgium, France, Germany, the Netherlands, Sweden and Russia. When they were first built, the frontiers, foothills and foreign ground which they crossed was colonized with ease, not only by wild seed carried up or down stream by water or boats, but also by escapees from the cargoes of corn, flour, fruit and so on. The result – a diversity of vegetation which has seen the unrestricted passage of plants for over one hundred years. Today, a canal bank may support anything from Russian comfrey, to Himalayan balsam, Japanese knotgrass or Deptford pink.

In Germany the canals link up the four large navigable rivers – the Elbe, Oder, Weser and Rhine – and connections are now possible between the Baltic and North Sea. In France, canals link the Mediterranean with the Bay of Biscay; in England the North Sea is linked to the Irish sea. In all cases, the continual passage of boats provides more than enough scope for exploitation by plants and animals. Populations of aquatic plants and invertebrates have been carried from country to country, setting up new colonies and establishing themselves on alien soil. Some species from distant places have met up, reproduced and produced hybrids with different characteristics from either parent. Others have adapted to the canal environment and remained unchanged – freshwater fish have moved in from some rivers and there are even cases of marine fish moving into canals from the sea.

HITCHING A RIDE

The canal systems of Europe have been responsible for dispersing several species of plant and animal to new areas. The boat's plying up and down the waterways help to spread the fruits of plants and also break up others – whipped to pieces by the propeller blades – which grow into completely new plants vegetatively.

Several species of plant have become a nuisance in canals, bunging up waterways and hindering transport. A tiny floating plant barely 4mm across, the infamous duckweed, can cause considerable chaos. During the summer months, these plants reproduce at an alarming rate, building up to such high levels that the water becomes covered with a uniform green mantle. This can pose a threat to those plants growing beneath them in

PLANT SUCCESSION

Terrestrialization is a rather fancy word used to describe the way in which all watery habitats have the tendency to dry up and eventually become dry land. The process may take many years but can be seen taking place along the edges of many watery habitats – particularly where the water is still and shallow. The culprits that cause this change of habitat are the plants.

As silting up occurs, the water will eventually become shallow enough at the edges to allow plants with floating leaves – water lilies for example – to become established. After further silting and stabilization, emergent vegetation, such as reeds

The flat wetland of the Camargue which lies between the two arms of the Rhone becomes white with water crowfoot in the spring, almost forming a natural monoculture. Rushes also abound, providing an ideal habitat for marsh frogs whose perpetual croaking fills the air at this time of year.

The Camargue is not all like this – much of it has been 'improved,' to make way for green paddy fields or white salt pans, both of which are colonized by opportunist wildlife.

and rushes, will grow. These encourage a great accumulation of deposits and also add to the debris themselves with dead leaves, stems and roots to bring the silt level up further still.

Eventually, the level will rise enough to allow water-loving plants like sundews and cotton grass to become established, forming a bog. Trees such as willows and alder, mosses and other plants can also be found growing in such water-logged conditions, all helping in the gradual transformation that turns open water into dry land.

Most shallow freshwater bodies exhibit plant succession in this way, hence the need for conservation groups to dredge ponds, lakes and streams to prevent them becoming strangled out of existence by their plant life.

the water since the duckweeds block the sun's rays, thereby stopping vital photosynthesis. Algae and blanket weed may also become a nuisance, along with other submerged plants like the crowfoots, hornworts and milfoils which grow quickly and form dense clogging patches.

Pondweeds grow very successfully in canals. There are at least 20 species but one of the most notable to have been transported via canals is Canadian pondweed. A native of North America, this species spread throughout the European canal network during the nineteenth century.

Students of biology and those with aquariums will be familiar with Canadian pondweed as the plant sold for aerating water tanks with bubbles of oxygen. Like many aquatic plants, it reproduces vegetatively which is the key to its success. Little bits of the plant will grow into complete plants by themselves without the need for sexual reproduction. Other examples of canal-transported plants include the floating water plantain, orange balsam (another North American species now found along rivers and canals throughout Northern Europe), slender rush and the water fern, Azolla – another American species which has thrived in the waters of Western Europe, forming dense mats of vegetation up to 2 kilometres long.

It is not only plants which have been inadvertently spread by canals. One American crustacean, the tiny shrimp *Crangonyx pseudogracilis*, is known to have first entered Northern Europe as part of a shipment of aquatic plants destined for home aquariums. The first example of this species in Europe appeared out of a tap in a London suburb in the early 1930s and later further examples turned up in Scotland, thought to have been transported across the Atlantic hidden in imported timber. The practice of seasoning timbers in ponds may have given this crustacean the opportunity to conceal itself deep inside the fissures in the damp wood. By the 1950s it had spread through much of Northern Europe.

Above: *The beautiful demoiselle is one of the commonest damselflies to be seen around aquatic habitats in Europe. The male is a bright blue; the female, seen here, an irridescent brown. These insects are found along slow-moving streams, pools and canals. Unlike their more powerful relatives, the dragonflies, they have a very delicate flight and slower wing beat.*

CANAL INVERTEBRATES

Canals can be alive with invertebrates, particularly insects. There are many which live their whole lives under water whilst others merely use water for breeding. The latter may sometimes be found around the garden and house if you live near a water source, but they never venture far from their breeding sites. Some prove to be unwelcome visitors – mosquitoes for example – but others, like the colourful dragonflies, can be fascinating to watch.

Of course, not all aquatic invertebrates are insects. The crustaceans are very numerous too. These are the freshwater relatives of the crabs and prawns that are so well represented in salt water. Three other groups may also be seen. These are the aquatic mites and spiders, the molluscs (principally aquatic snails), and the 'worms' that live in the mud of the canal bottom. All serve as a vital source of food for larger animals such as fish and birds; an essential part of the food web and chain which nourishes their predators. The further down the food web they are, the more numerous they tend to be since larger quantities of them have to be eaten in order

***Left:** Water lilies, pondweeds and duckweeds all have circular, oval or elongated leaves which float on the surface of the water. They act essentially as solar panels, capturing the sun's energy with their chlorophyll pigments and making sugars in the process of photosynthesis. These leaves can completely blanket the water surface, turning small ponds, dykes and ditches green with their foliage.*

***Right:** The fire salamander is boldly coloured yellow and black – a warning to predators of the poisonous secretions in its skin. These slow-moving amphibians occur widely in Western Europe and can sometimes be found living in damp cellars, emerging to feed after rainstorms.*

A GUIDE TO SOME AQUATIC INSECTS

There are plenty of insects associated with water – many a fisherman's fly is modelled on the different ones which hatch out through the year. Some live on the surface, others spend their lives submerged amongst the water weeds. These latter creatures still have to breathe air and many return to the surface regularly to trap a further supply in the fine hairs on their bodies. The best way of studying aquatic insects is to get your nose down to the water's edge or sample them by means of a fine net.

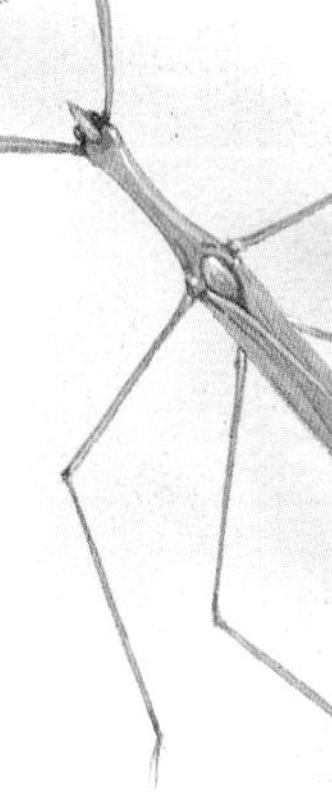

***Water boatman.** These are submerged true bugs which have a highly-modified pair of back legs. These are used for swimming, rather like oars, and are very effective in propelling the insect around in the water. (Length: 16mm)*

***Whirligig beetle.** These beetles are nearly always found in groups. They whizz relentlessly around on the surface and can be very common. (Length: 6mm)*

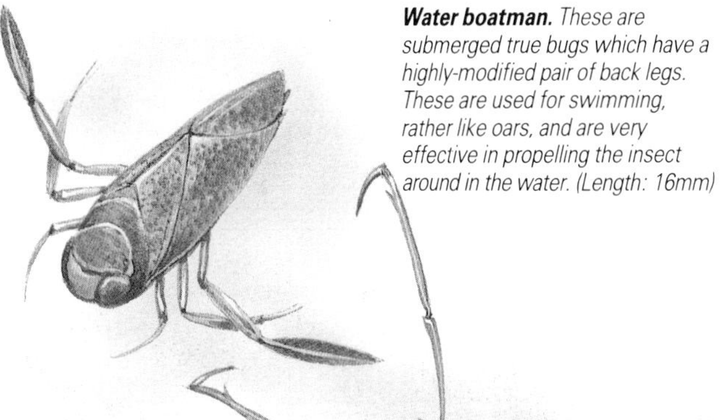

***Water stick insect.** Similar in appearance to the terrestrial stick insect, this creature in fact belongs to a completely different order – that of the true bugs. It has a long breathing siphon which may be mistaken for a tail. (Length: 50mm)*

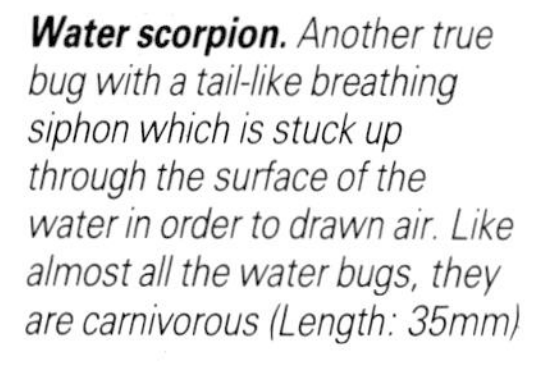

***Water scorpion.** Another true bug with a tail-like breathing siphon which is stuck up through the surface of the water in order to drawn air. Like almost all the water bugs, they are carnivorous (Length: 35mm)*

***Great diving beetle.** A fierce carnivore which will eat tadpoles, newts and small fish, as will its predatory larvae. It is common in village ponds, canals and dykes. (Length: 35mm)*

to fulfil the predator's daily energy quota.

The best way to study invertebrates in old canals is to lie face down on the towpath and study life in the water from 3 or 4 cms above. The still, clear water will soon come alive with lots of tiny creatures: snails and leeches slithering over leaves, predatory dragonfly nymphs lurking in the weeds waiting for passing insect prey and the tiny red water mites swimming to and fro laying their mass of eggs. Other curiosities worth looking out for are water scorpions and water stick insects, both named after their likeness to other land-dwelling invertebrates. They both breathe through long tail-like appendages which are pushed up through the water's surface.

Cast your eye over any stretch of freshwater and you will usually find at least one pond skater. Like other surface-dwelling bugs such as the water measurer, pond skaters can literally walk across the water, their feet supported by the surface tension. Other predatory bugs such as the water boatmen, live just beneath the surface, propelling themselves around with their large hind legs which are used like oars. They

DRAGONFLY METAMORPHOSIS

The metamorphosis of the dragonfly takes about two years. The early stages are spent in the water, the adult insect being the only stage that lives in the air. Both nymphs and adults are carnivores and will predate on all sorts of insects and invertebrates.

The eggs are often laid on submerged vegetation, the female making a fine incision in which to lay them. Other species plunge their abdomens deep into wet sand at the edges of streams and lay the eggs there. The nymphs hatch a few days later.

The nymphs lurk amongst vegetation waiting to pounce on some aquatic insect or small fish. It shoots out its mouthparts, or 'mask', at its prey. They are hinged back under the head when not in use.

Prior to hatching into the adult dragonfly, the fully-grown nymph crawls out of the water on to a plant stem. Its larval skin splits down the back to reveal the perfect adult which throws itself backwards in an effort to wriggle free.

The emerged dragonfly will remain on the plant stem until its wings have dried and hardened. It then flies off to mate and begin the whole process once again. The adults live for about a month.

Dragonflies are powerful fliers and reach speeds of up to 30kph. Their wings make a rasping noise when in flight. Some species are migratory. They have bright colours, large compound eyes and can be roughly divided into two types – hawkers, which patrol the waterways looking for insect prey, and darters that rest on vegetation or some other perch and sally forth to catch passing victims.

Left: *Pond skaters are blown into great congregations on the water's surface by the wind. They are related to the ubiquitous aphids and cicadas and have successfully mastered the art of skimming over the water by using the surface tension to buoy them up. They are carnivores, always on the look out for small insects which fall or are blown into the water.*

Below left: *Caddis fly cases can often be found under submerged stones and rocks. The larvae of these flying insects live underwater, spinning a protective coat of silk into which they incorporate tiny sand grains and other debris. Safely inside with only its head protruding, the larva can then spin a net to catch its food of sieved plant and animals remains. The adult flies are often attracted to lighted windows. They are brown with long antennae.*

frequently return to the surface to replenish their supply of oxygen, their undersides being covered with fine hairs which are used to trap air before they submerge, giving them a silvery appearance. There may also be aggregations of whirligig beetles. These grow up to 6 millimetres in length and can often be seen darting to and fro across the water in groups or swimming energetically in tight circles. The speed at which they travel is quite staggering, moving at roughly a metre per second, though only over minute distances.

Spiders and their closely related cousins, the mites, are equally at home on or in the water. One species, the water spider, both lives and breeds underwater. Like the bugs, it uses the fine covering of hairs on its abdomen to collect air each time it surfaces. It then uses this to stock its airtight nest of silk which is attached to vegetation. When sufficient air has been collected in its 'diving bell', the female lays up to 100 eggs in the nest which develop underwater. Another spider to look out for is the raft spider which lives by the water's edge. It hunts for small aquatic prey by walking across the water, using the surface tension for support. Only a few millimetres in length, it has been known to catch even small fish.

Aquatic insects with flying adult forms undergo a complete metamorphosis, rather like that of the butterfly or moth. The change from water-dwelling larva to airborne adult is usually achieved by the fully grown nymph which crawls up on to a reed stem or any other convenient surface that protrudes from the water. The adult form then emerges, leaving behind its nymphal case. You may come across hundreds of these cases, indicating the site where thousands of mayflies have just emerged.

There are five separate groups of insect which undergo this change: the dragonflies, damselflies, alderflies, caddis flies and stoneflies. Perhaps the most obvious insects from this group are the damsel and dragonflies which are regularly seen cavorting above the water. The beautiful damselflies are often found in large groups. Their bright metallic colours stand out in the sunlight against the leaves of alder and willow where they endlessly court and chase their insect prey. Watch out for them fluttering delicately over aquatic vegetation such as watercress and mints, alighting on strategically placed flowers ready to launch their attacks on small insects. Like the larger dragonflies, they are predatory insects, often catching their prey in full flight with the aid of their bristly forelegs.

Dragonflies fall into two separate types – hawkers and darters. Hawkers tend to be slightly larger with longer bodies and a greater wingspan. Both larval and adult dragonflies are fiercely predatory, tirelessly patrolling a stretch of river or canal searching for their insect prey which sometimes includes their smaller relative the damselfly. Darters tend to spend more time resting on the waterside vegetation, sallying forth only to catch potential prey or in their search for a mate.

AQUATIC INVERTEBRATES	
Alderfly	Pond skater
Caddis fly	Pond snails
Crayfish	Scorpion fly
Daddy long legs (cranefly)	Swan mussel
Damselfly	Water boatman
Dragonflies	Water measurer
Great diving beetle	Water mite
Leeches	Water scorpion
Mayfly	Water spider
Mosquito	Water stick insect

Contrary to belief, dragonflies are not harmful – unless, of course, you happen to be a fly! They do not sting or bite, but are immensely efficient hunters with their powerful jaws, four large wings and clawed forelegs. They are equipped with perhaps the largest compound eyes of any insect, with thousands of small individually-lensed units called 'ommatidia' making up each side of the eye; rather like looking at a few thousand television sets all at the same time. This helps them to detect directional movement (essential in their hunt for prey) rather than detail. Their sight is matched only by the hoverflies.

Dragonflies lay their eggs in a number of different ways. Some sit on the water's surface, plunging their abdomens deep into the water to place a single egg in a slit in the vegetation. Others will rise rapidly up and down over shallow water, placing eggs in the damp sand each time their abdomens hit the water. Female craneflies exhibit similar behaviour when they probe wet mosses when laying their eggs.

Unfortunately both dragonflies and damselflies have disappeared at an alarming rate from many waterways over the last decade, some becoming extinct in localities where they were once common. The problem is habitat destruction – the filling in of ponds and canals, pollution and enrichment with agricultural chemicals. For this reason, garden ponds can often be a welcome refuge.

Studying the vegetation along the water's edge may also reveal some interesting snails such as the great pond and impressive ramshorn snails. Like their terrestrial relatives, the aquatic snails creep slowly over vegetation, rasping away at the plant cells. Other molluscs to look out for are the wandering snail, a pollution-tolerant species, the river pea mussel and the large swan mussel which can reach up to 23 cms in length and live for up to eleven years in the muddy bottom of a canal or river.

One of the commonest crustaceans in the water is the freshwater shrimp, *Gammarus*. Mating couples can be seen swimming around in tandem, looking rather like fleas with their laterally-flattened bodies. They feed on decomposing vegetation. Canals may also harbour large numbers of an aquatic woodlouse called the water slater. They look rather like their land-living relatives and can be found even in poorly oxygenated or polluted canals, breathing through special gills on the underside of their bodies. Freshwater crayfish, smaller relatives of the sea lobster, will only be found in clean waters. Their presence frequently indicates a pollution-free aquatic environment.

Above: *Yellow iris or yellow flag is a very common sight in wet areas. It can often be found growing alongside water dropwort along water courses, streams and ditches. Its flowers are bright and distinctive, especially when seen in damp woods. Like the garden irises, it has an underground rhizome which produces side shoots as it gets larger. Each year new leaves and flowers are produced and the plants can completely choke a small stream.*

BANKSIDE PLANTS

Apart from free-floating aquatic plants (like duckweed) and submerged plants (like the pond-weeds), there is another group of plants found at the water's edge. These grow with their roots firmly entrenched in the mud of the canal bottom but have aerial leaves and flowers. Perhaps the most familiar of these to be seen along canals are the reeds. Stands of reedmace may often be found along the water margin. There are two sorts, greater and lesser, both of which are popularly and incorrectly known as bulrushes. Their distinctive brown fruiting spikes soon turn

PLANTS	
Creeping buttercup	Teasel
Greater willowherb	Watercress
Lesser celandine	Water crowfoot
Lesser spearwort	Water dropwort
Marsh marigold	Water fern
Meadow buttercup	Water parsnip
Meadowsweet	Water plantain
Golden saxifrage	Water milfoil
Reedmaces	Water starwort
Rushes	Yellow water lily

TYPES OF AQUATIC VEGETATION

A wealth of plant life can be seen at the water's edge. There are those which like to grow near the water but not in it; those that are rooted in the mud but have emergent leaves; and others which are submerged apart from a few floating leaves. Some species grow completely underwater, whilst othes waft around on the water's surface and have no permanent base. If left alone, their colonization is often so successful that eventually no water can be seen.

Aquatic plants provide cover for numerous insects, and often form dense waterside thickets that are an ideal refuge for nesting birds. Watch the vegetation for the comings and goings of insects and birds next time you take a walk along a towpath.

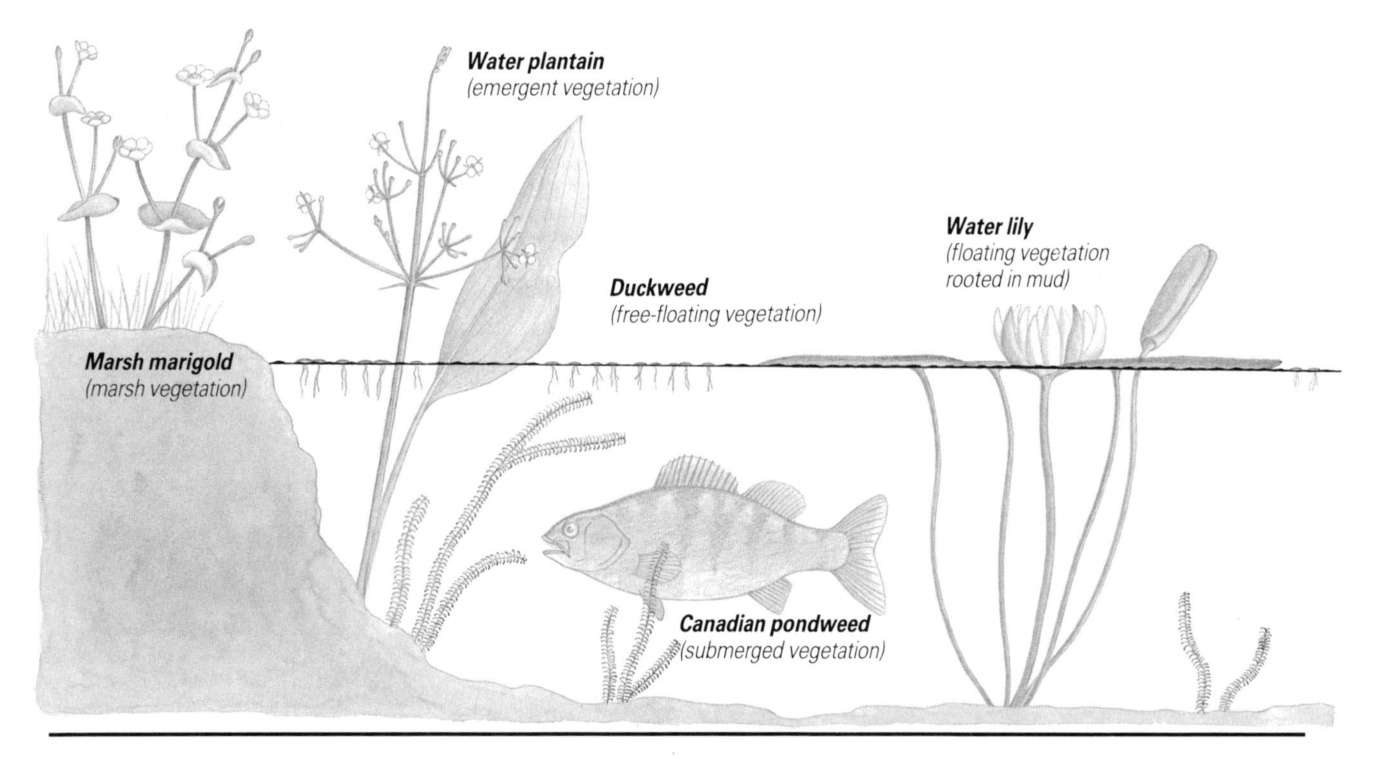

into a mass of fluffy seeds which take many months to disperse, some remaining on the spike until the following spring. Branched bur reed may also be seen in shallow water or mud, prefering the slow moving water of canals to the hustle and bustle of larger rivers. Common rushes are another familiar sight, forming impenetrable masses at the water's edge, their flowers born as tiny swellings towards the end of the narrow stems.

Emerging from the water's edge, water plantain can often be seen standing up to 2 metres high. Other plants to look out for include the arrow-like leaves of arrowhead thrusting up from the water's surface, the white and curiously hairy flowers of the bog bean, yellow and white water lilies and the delicate flowers of milfoil.

The damp zone of the canal bank harbours further wetland flowering plants which make colourful displays during the spring and summer. Stands of greater willowherb, purple loosestrife and yellow iris are typical. Smaller plants include the bright yellow marsh marigold or kingcup which is often found growing in great clumps, water mints and the blue flowers of brooklime. Tall stands of water dropwort often smother large areas, looking rather like cow

Above: *Reedmaces are very common on the banks of canals, gravel pits and reservoirs. They have light fluffy seeds which disperse gradually from the tall brown fruiting spikes. Reedmaces are important plants since they colonize shallows, an essential part of the succession of plants that occurs in waterways, and also provide dense thickets for nesting birds.*

Left: Purple loosestrife is a spectacular plant which thrives near water. Low-lying meadows are often ablaze with its flowers and prime specimens can often be found leaning over the water at the edges of canals and reservoirs. The many small blooms are often visited by honey-bees for nectar.

Right: Mallards are the ancestors of many domestic ducks and are still one of the commonest wild species in the world. They breed prolifically, laying 10 or 12 eggs in a nest made in vegetation by the water's edge. The male can be easily recognised by his irridescent green head and white neck for most of the year, although in eclipse plumage he looks almost identical to the female at the back of the picture. Mallards readily colonize a variety of man-made habitats, including canals, reservoirs, boating pools, and town ponds and lakes.

parsley from a distance.

The trees that colonize wet areas are the willows – a large group which includes the sallow and the riverine alder which has a reddish bark. Frequently you will come across waterway edges entirely colonized by these trees which are always insect-rich.

CANAL FISH

Man-made waterways contain plenty of fish but the species found there will depend on the type of water (whether fast or slow moving) and the relative level of pollution. So good can certain fish be as indicators of the health of any stretch of water that some local authorities, anxious to keep waterways free of pollution, will sample fish for clues. The most likely species to be found in canals are roach and perch, though bream, carp, tench, pike, gudgeon, ruffe, chubb and sticklebacks may also be found. Sticklebacks are usually found in cleaner water and are an indication of such.

Fish consume a wide range of food, some preferring only algae and plankton, others feeding solely on insect larvae and a few (the pike being the best example) taking the occasional amphibian, water vole or even duckling, as well as other fish. But although food is often plentiful in canals, many fish require faster-moving or cleaner water in which to live.

The perch is a common fish in most slow-moving rivers or canals, and is possibly one of the most beautiful freshwater species. Its ventral fins are bright orange edged with red, the greeny grey dorsal being supported by sharp spines. The hump-backed body is sometimes markedly striped dark blue-green. These fish can often be found in shoals feeding on aquatic larvae and even their own fry. The roach, on the other hand, is an omnivore, devouring plankton as well as insects and their larvae. During the breeding season, the females of the species shed several thousand pink eggs so that spawning grounds often become covered in carpets of spawn.

Other fish may also be found in canals which have either been deliberately introduced or moved in from rivers or the sea. Carp and goldfish are two examples of introduced fish which are now known to be breeding successfully in some canals in Northern Europe. Permanent colonies of goldfish – even tropical fish – may build up as a result of unwanted pets being thrown unceremoniously into nearby waterways.

The Crucian carp – a native of the Far East – has also been released and serves some use as a waterweed eater! In fact, research in Holland has shown that the use of grass carp, an introduced species from China, can be successfully used as a grazer of clogged-up canal and river systems. There, it has reputedly decreased the cost of waterway maintentance by 40%.

FISH	
Bleak	Pike
Common bream	Roach
Common carp	Rudd
Gudgeon	Stone loach
Perch	Tench

CANAL BIRDS

Old canal banks overgrown with sycamore, silver birch, ash, willow and alder may be expected to have the same sort of birdlife as that along the scrubby thickets of disused railway lines. During the spring and summer, willow warblers sing their descending scales, linnets twitter from exposed bushes, yellowhammers call out from the overhead branches and meadow pipits may be found nesting in suitable places on the ground. More noticeable, however, are the waterbirds, namely the coot, moorhen, dipper, heron, kingfisher and the familiar mute swan and mallard.

Dippers are quite remarkable birds since they can walk or run underwater, collecting mouthsful of aquatic insects as well as caddis fly larvae, tadpoles and worms. They can often be found feeding near sluices and waterfalls where they stand in the fast-flowing water, or on rocks in the thick of the main current.

The blue flash of the kingfisher is quite unmistakable as it skims past, flying low over the water before diving in to catch its prey of small fish. These birds make holes up to a metre deep in sandy banks and there lay their six or seven eggs. The young are fed small fish such as sticklebacks and, like their parents, swallow

THE SWAN'S SONG

The mute swan, which is the most common swan on waterways throughout Europe, has been unable to live a life free from the effects of man. If not affected by river or canal pollution, swans (along with plenty of other water birds) frequently fall prey to lengths of discarded fishing line which become caught up around their legs and necks. But perhaps their most pityful affliction is lead poisoning which occurs as a result of the birds swallowing discarded lead fishing weights. Although urban birds tend to do better than their country counterparts as far as food is concerned (since there are more people to feed them), they tend to have higher concentrations of lead in their blood. Poisoning can result if the level becomes too high, and affected swans can be recognised by the 'broken neck syndrome', since the lead breaks down their neck muscles leaving them with a permanent stoop. Eventually, the birds die from starvation because they can no longer use their neck muscles for swallowing. However, the introduction of a new kind of fishing weight (made from a tungsten polymer) will hopefully reduce the number of deaths from poisoning and ensure the continued presence of these beautiful birds along our waterways.

Mute swans usually pair for life and return to their nesting grounds in the spring, searching out lakes, ponds and other open bodies of water well

The mute swan is so called because it makes no sound whilst flying, unlike other European species. Up to seven cygnets may fledge from the domed nest of plant material, their dense grey-brown down soon being replaced by the snowy whiteness of the adult feathers. The parent birds will protect their young fiercely, hissing and flapping their wings at any intruder.

bordered by reeds. Despite their name, these swans snort, hiss and can produce a quiet trumpeting noise. The legendary song said to be made by dying birds has, however, no foundation in fact.

Whooper and Bewick's swans may also be seen on and around rivers, lakes and reservoirs. The two are very similar in appearance, although whooper swans can be recognised by their larger size and the greater area of yellow on their bills.

Left: *Life for swans in towns and cities has become much better. They are fed more regularly than those in the country, with bread and other food given by man. This has the effect of allowing them to lay heavier eggs than their counterparts in the countryside. The chicks also tend to do better and there is a greater rate of survival, leading to more cygnets being seen with their parents in urban habitats than in the countryside.*

Below: *Flocks of Canada geese are seen frequently on inland waterways. Originally introduced from North America, they have been enormously successful to the extent of becoming a nuisance in some urban and rural areas.*

them head first so that the scales do not stick in their throats. Though more usually seen in flight, these birds may also be seen perching or hovering near the water before diving in to catch small fish such as sticklebacks, minnows or gudgeon, though water beetles, dragonfly nymphs and other invertebrates may also be taken.

An expert hunter, the impressive heron is also a familiar sight along the edges of canals and streams. A bird of seemingly endless patience, it will stand motionless or stalk through the shallows ready to catch its prey of fish, frogs or water voles. Herons are quite distinguished in appearance with their long yellow bill, black crest and grey upper parts. They can be easily identified in flight, their heads drawn back and their long yellow legs trailing behind.

Perhaps the most familiar waterbirds, however, are the ducks and swans. Mallards in particular have learnt to live alongside man in the ponds of his town parks and are often tame enough to take a crust from the hand. The male is easily recognised with his brilliant green head and white collar, although he becomes almost identical to the dull brown female during the autumn months. When in this eclipse plumage,

CANAL BIRDS	
Coot	Moorhen
Dipper	Mute swan
Heron	Pied wagtail
Kingfisher	Swallow
Mallard	Yellow wagtail

his yellow bill is the only real means of telling the two sexes apart. These ducks can also be found feeding on farmland, often several miles from the nearest stretch of water, supplementing their normal diet of water plants with grain and weeds.

Moorhens and coots are often confused. Both birds are omnivores found wherever there is water with sufficient vegetation to supply them with cover. Moorhens are easily identified by their bright red bills and foreheads. They display some unusual characteristics, not least the fact that when alarmed they will sink below the water leaving only their bill protruding like a periscope. Coots have a similar blue black plumage with a white bill and forehead. They have distinctive feet, having long toes with scalloped edges that help them to run across the water before taking flight. Like moorhens, coots are fiercely territorial, fighting off intruders before taking refuge amongst the waterside vegetation.

RESERVOIRS AND GRAVEL PITS

Man has dramatically altered the landscape in Europe. Artificial lakes have been created through gravel pits, mineral extraction, clay pits and peat digging. Streams and rivers have been dammed and scenic valleys flooded to create vast reservoirs that ring most major towns and cities throughout Europe.

Everyone should find they have at least one reservoir or gravel pit somewhere nearby. Such habitats provide amenities for humans as well as wildlife – places to walk and take fresh air, places to fish and places for water sports such as sailing, power boating and wind surfing, but as habitats to study plants and animals, they offer a wealth of opportunities. Reservoirs are often more accessible for nature watchers than gravel pits, but wildlife knows no bounds. Some species colonize both habitats; others stick to just one.

The type of water found in gravel pits and reservoirs is also influenced by man. It tends to be mineral-rich (eutrophic) which helps wildlife to thrive. This is in direct contrast to, for instance, highland lakes where the nutrient content of the water is poor and the wildlife potential therefore less. There are some species of birds, however, that are only ever found in such (oligotrophic) waters. The great northern diver is one good example, proving yet again that the vast majority of wildlife on our doorsteps consists of those species that are able to adapt and take advantage of the new habitats offered by man and his environment.

As with canals, the types of habitat available to wildlife in and around these wetland bodies include more than just the water itself. Gravel and sandy banks, marshes, reedbeds, shallows, islands, marginal thickets and mudflats are just a few of the others. Plants are not particularly fussy about where they grow and often flourish amongst the mess of discarded industrial machinery, frequently succeeding in concealing it completely after just a few years. What was once a thriving industrial site can often be reabsorbed into the landscape, no trace of its former role being evident to the passer-by.

FAVOURS FROM FACTORIES

It is a well established fact that the warm water of power stations encourages plant and animal growth in the immediate area. Power stations need to draw off water for cooling purposes and this is returned warm. Too much heat can kill, but a gentle outflow that is just a few degrees warmer than normal can create a special environment which encourages the growth of algae and plankton, thereby attracting small invertebrates, fish and birds.

Sea horses, guppies and other tropical fish have all been found breeding in artificially-warmed waters near to power stations and factories, and a Mediterranean bladder snail normally found in warm greenhouse tanks has become established in at least one river in Northern Britain. Factories in close proximity to urban gravel pits can also be a good thing. The winter temperature of the water is often higher than that of other bodies of water, a fact that encourages wintering birds such as the bittern to stay. The warmer water also induces the early hatching of fish, proving yet again that living in the urban environment can have distinct advantages.

A disused canal running through an urban area offers many refuges for wildlife. The proximity of power stations, electricity pylons, sewage pipes and housing estates do nothing to deter colonization by dragonflies and other aquatic insects, numerous plants such as the reedmace that can be seen encroaching towards the centre.

Sometimes the warmth given off by the urban environment boosts both plant and animal growth – provided, of course, that there are no toxic pollutants.

WILDLIFE AROUND RESERVOIRS

Wildfowl have never had it so good as on the man-made gravel pits and reservoirs built in the last few decades. Thousands of hectares of open water have been provided for feeding and resting places and, of course, it is not only the birds that have taken advantage of these new habitats.

Around their edges dense reed beds spring up, banks of yellow iris bring colour to the waterside and carpets of water crowfoot cover the water with tiny white flowers. Aquatic insects of all kinds provide ample food for birds and amphibians alike and, in the case of dragonflies, may be seen sporting their bright colours as they dance over the water surface. The fishing is good too, as a result of so many introductions by man in the past, and amongst the lush vegetation mink and other mammals hunt their prey.

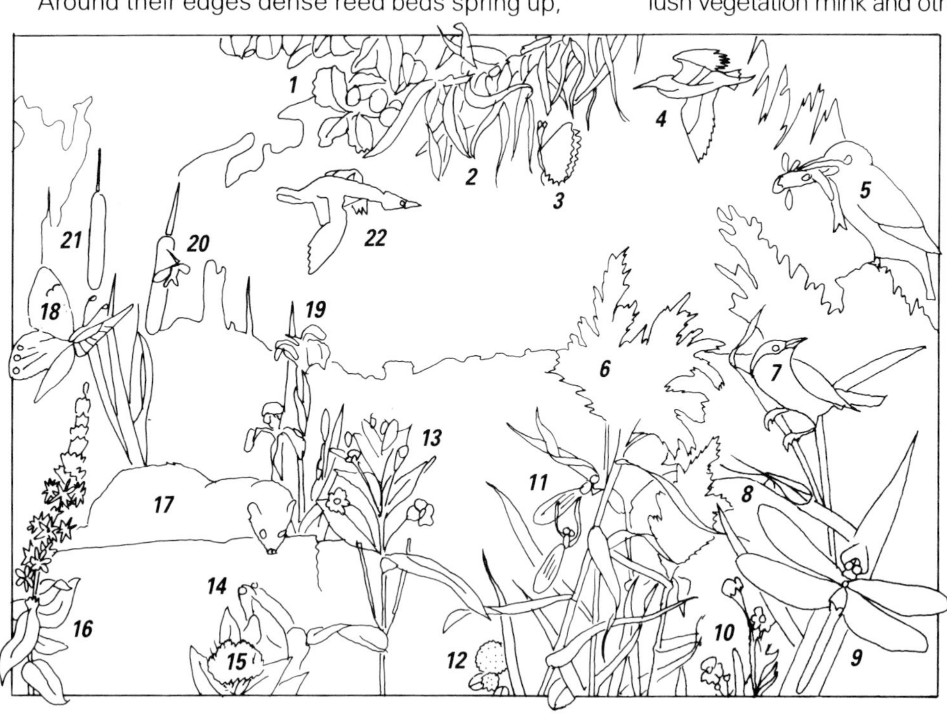

1 *Alder*
2 *Willow*
3 *Comma butterfly*
4 *Heron*
5 *Kingfisher*
6 *Common reed*
7 *Reed warbler*
8 *Caddis fly*
9 *Common aeshna dragonfly*
10 *Spearwort*
11 *Red damselfly*
12 *Mint*
13 *Greater willowherb*
14 *Common frog*
15 *Hemp agrimony*
16 *Purple loosestrife*
17 *Mink*
18 *Clouded yellow butterfly*
19 *Yellow iris*
20 *Cranefly*
21 *Greater reedmace*
22 *Great crested grebe*

BIRDS OF THE OPEN WATER

Throughout Europe the increased incidence of reservoirs and dams has provided wildfowl with many more watery habitats to use. It is, therefore, not surprising that some birds have taken advantage of them and done extremely well. Some, like the geese, have undergone population explosions to the extent that certain species are now regarded as pests. Such birds use the large expanses of water for shelter and make regular sorties to nearby agricultural land where they feed on the fresh green vegetation conveniently laid out for them in nice orderly rows. In general, the effect of reservoirs has been to increase the number of bird species found in any one locality, particularly migrant wildfowl. This success is best illustrated by the migrant ducks, geese and swans.

At the turn of the century, migrating ducks flying high over the countryside would have seen only a few open bodies of water – the earlier reservoirs of the Industrial Revolution and some even older mill ponds and lakes. Today, their horizons have opened considerably. Below them there are a myriad of waterways, large expanses

of blue offering food, shelter and a relatively safe place to rest before continuing their journey. The choice is theirs.

However, the provision of these artificial lakes is not enough to successfully (and continually) attract wildfowl. They require more than this. Food and shelter are also of paramount importance, especially if successful breeding is to take place, and although much of the food may make itself available naturally (through colonization by aquatic plants, insects and fish), much can also be done by man to encourage the birds' success. One good example is the encouragement of extraction companies to make pit edges wavy rather than straight, a conservation strategy devised by the late Dr Jeffery Harrison. This increases the number of possible nesting sites for breeding ducks and other birds, and reduces the chances of them abandoning their nests when disturbed by unwelcome movement on the horizons. Moreover, spits and sheltered corners have also been created, providing the birds with secluded places where they can gather to moult and preen.

Another worthwhile provision is one of making islands for breeding wildfowl. Prowling foxes and other predators can drastically reduce the numbers of many species by taking the adults, eggs and young of those birds which nest along the banks of reservoirs and similar bodies of water. By creating artificial islands only accessible by the birds themselves, fatalities can be considerably reduced. After various attempts, it was found that floating islands made from planks wired to oil drums were the best solution. These were then covered with soil and gravel, anchored to the pit bottom and left for the birds to discover. Such islands have proved to be very successful wherever they have been established. Two birds in particular have taken to them – the Canada goose and the common tern. In fact, the increasing spread of the latter bird inland is believed to be due to its use of gravel pits for feeding and breeding.

A final, and equally successful, strategy put forward by Dr Harrison was the planting of specific food plants for wildfowl around the edges of gravel pits. By analysing the crop contents of a few birds, conservationists were able to produce a list of trees and shrubs on which the birds had been feeding. A planting scheme was then instigated which has resulted in prodigious amounts of the right seed being produced each year and this has, as expected,

Below: *On many reservoirs, nature has to vie for attention with water sports – sailing, surfing and fishing. However, many such places have quiet corners which are set aside for conservation. These nature reserves provide excellent opportunities to study freshwater life, from invertebrates in the mud to the many wildfowl that are attracted to these sanctuaries.*

Above: *The Carolina duck is, as its name suggests, a North American import. It is also known as the wood duck and is a familiar ornamental species with its attractive plumage. Very few pairs have succeeded in surviving in the wild but escapees can be seen occasionally on reservoirs throughout Europe. Other escapees from wildfowl farms which may be seen include black swans, chestnut teal, Chiloe wigeon, marabou storks, greater and Chilean flamingoes, and sarus cranes.*

acted like a magnet to the birds.

At least four species of wildfowl have increased in number and distribution thanks to the attractiveness of reservoirs and gravel pits. These are the great crested grebe, tufted duck, pochard and smew. All four require deep water in which to dive for vegetation and this is amply provided for, especially with deep water aggregate extraction.

The sight of great crested grebes on reservoirs used to be a rare sight. Today there is ample opportunity to see pairs of these handsome birds on reservoirs and other open bodies of water throughout Europe. They regularly breed and feed well into the urban environment and have successfully exploited the large open spaces of water provided by man's mining activities.

During the spring breeding season, the parent birds undertake an elaborate courtship ritual. Both birds will dance across the surface of the water, shaking their heads and spreading their wings in a distinctive display, designed to secure a mate. The birds then make a large and untidy nest of waterweed and debris which is either situated at the water's edge or anchored to reeds or other plants. The parents take turns at brooding the eggs and may be seen carrying the resulting chicks around on their backs. The little grebe or dabchick may also be seen on reservoirs, diving gracefully for small fish and water insects.

The tufted duck is another diving bird that is common on waterways throughout Europe. These birds have progressed from being relatively rare at the beginning of the century to being one of our commonest diving ducks. Easily recognised with their white flanks contrasting sharply with the rest of their dark plumage, tufted ducks are gregarious birds which can

RESERVOIR BIRDS	
Barnacle geese	Kingfisher
Black-headed gull	Little grebe (dabchick)
Canada geese	Mallard
Coot	Moorhen
Brent geese	Pink-footed geese
Gadwall	Pochard
Great crested grebe	Teal
Greylag geese	Tufted duck
Heron	White-fronted geese

A GUIDE TO SOME FRESHWATER BIRDS

Reservoirs are excellent places to study water birds. A great variety of species can be seen, although identification is not always easy since the birds vary not only from species to species but with age, sex and time of year.

When visiting such a habitat, a pair of binoculars is an almost indispensable piece of equipment, allowing you to get close-up views of waterfowl without the need to disturb them – and getting close is not always possible anyway, particularly if they are swimming about in the centre of a vast stretch of water. For more information on binoculars, see page 157.

Moorhen. *A common sight on many inland waterways, the moorhen can be distinguished from the similar coot by its red, rather than white, forehead. Nests are made at the water's edge and may contain up to eleven young.*

Pochard. *These attractive ducks can be recognised by their greyish body and chestnut head in the male, which is dull brown in the female. Their numbers have increased as more and more reservoirs are made.*

Kingfisher. *An easy bird to identify with its flash of blue colour. These birds are often found in pairs close to their nesting area, along streams, rivers and canals where there are high banks.*

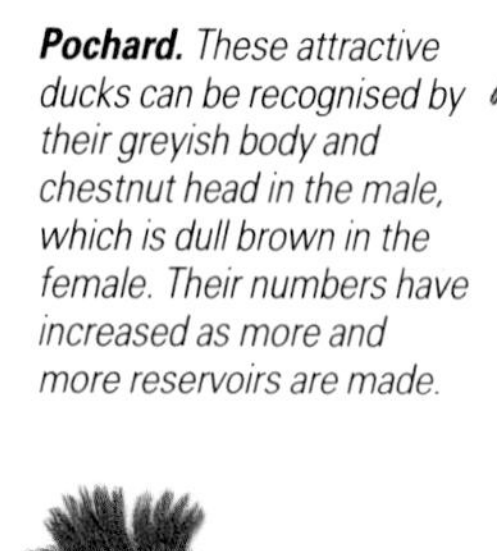

Great crested grebe. *These grebes were quite infrequent before man-made lakes provided them with the big open spaces of water on which they do so well. They are often seen in pairs and are well known for their extraordinary courtship behaviour.*

Right: *Shelducks are popular wildfowl which are frequently seen on park ponds as well as reservoirs. They are attractive birds with white, black and chestnut plumage which unusually is the same in both sexes. Shelducks normally nest near the coast, often using old rabbit burrows for shelter. They have also been known to nest in hollow trees and haystacks, occasionally venturing inland to woods and farmland to raise their brood of 10 or more chicks.*

Heron. *Herons can be seen standing motionless at the water's edge, ready to stealthily stab their prey of fish and frogs. Their greyish colour and fine feathers are an ideal camouflage, making them hard to spot amongst waterside vegetation. They roost on the branches of old trees and nest in heronries, some of which have been occupied for decades.*

often be seen nesting and feeding in quite large colonies. As with many waterfowl, their courtship ritual makes an interesting sight, the male tilting back his head and whistling softly while the female repeatedly dips her beak into the water and utters a loud growling call.

Smews and pochards are also increasingly common on reservoirs. Pochards can also be found in many town parks, the male easily distinguished by his handsome chestnut head. They nest close to the water, often in dense vegetation provided by reedbeds. Smew, on the other hand, are one of the few ducks to nest in holes in trees and were probably dependent on those left by the black woodpecker. In Northern Europe the latter bird is now quite a rarity and is completely absent from Britain, although in the mountainous forests of France, Germany and Scandinavia its familiar call may still be heard. To overcome the shortage of nesting holes, some success has been had using specially prepared boxes to encourage smews to nest.

Canada goose. *Canada geese are now a nuisance in many areas where they have bred prolifically on reservoirs and gravel pits. They feed on farmers' crops growing nearby and make regular flights over urban areas adjacent to their waterways.*

PROBLEM GEESE

Various species of geese may be seen on reservoirs and gravel pits, including Canada, Brent, barnacle and Greenland white-fronted. Canada geese have been around for a long time. As their name suggests, they originate from North America but have been seen in many North European parks since the seventeenth century when they were imported as ornamental species. Like many other geese, they use reservoirs and gravel pits as night roosts, breeding on islands and spits and grazing on the surrounding agricultural land much to the chagrin of farmers. Flying to and fro from their feeding grounds, they often pass over urban areas and can be seen in many town parks, fields and marshes. They are gregarious birds, frequently gathering in flocks of several hundred. They also nest in colonies, each bird lining a depression in the ground with grass and down, well hidden amongst the waterside vegetation. Migrating flocks can be seen flying in the familiar 'V' formation or forming a long line across the autumn sky.

OTHER VISITORS

Apart from waterfowl, reservoirs receive visits from a number of other birds looking for food and shelter. Birds of prey, including hen and marsh harriers, buzzards and long-eared owls, can be seen searching the water's edge for unsuspecting chicks, water voles and other small prey. The osprey, too, occasionally makes a rare visit to reservoirs and gravel pits on its passage to warmer climes, sometimes offering a unique view to bird-watchers as it plunges towards the water to snatch a fish swimming dangerously close to the surface.

Seabirds normally associated with coastal areas and marshes are now frequently found well inland, roosting and feeding on these open bodies of water. Many of the reservoirs are rich in fish and invertebrates on which these birds feed and it is not unusual to find flocks of over 10,000 black-headed gulls amassed on just one reservoir. Great skuas can also be found well inland, along with various species of tern – the Arctic, common and Sandwich – as well as other members of the gull family.

One quite regular seabird visitor is the cormorant, a slim diving bird more usually seen drying its wings on a convenient rock near the shore. In the reservoir setting, the dark forms of these birds can be recognised at a distance, often perching on dingies, buoys or other elevated surfaces alongside the familiar heron.

SHALLOWS AND SHORELINES

When reservoirs are created, much marginal land often becomes flooded to create marshes or areas of shallow water which offer further habitats for birds. In the marginal shallows the water may be warmer, encouraging a good growth of vegetation and a ready supply of food for shallow water feeders, including gulls and snipe. The shoveler and pochard are two others that can be found breeding and feeding with great success.

Above: *Well-fed mallards take up their typical resting position whilst standing on the ice at the water's edge. In the wild they would have a hard struggle finding food at such times of the year but, like the mute swans, they have benefitted from living close to man and surviving off the food so readily supplied to them in the urban environment.*

Similarly, the stony shorelines around the edges of many reservoirs and gravel pits are prime breeding sites for the ringed and little ringed plovers. Both birds like these exposed situations and make shallow depressions amongst the stones in which to lay their eggs. If disturbed, the nesting ringed plover will distract the enemy away from its nest and young by feigning a broken wing – a ploy also used by the curlew. Watch out for feeding parties of plovers running about at the shoreline, bobbing their heads up and down as they search for small invertebrates.

Gravel pits often support more habitats than simply the open water. Reedbeds frequently develop around their edges in the loose sandy soil and these attract warblers which move deftly between the reed stems accompanied by their

Above: *A juvenile herring gull stops for a rest on a sun-warmed drain cover in a busy car park. Such coastal species are frequent high street visitors in coastal towns and sometimes venture inland to lakes and reservoirs, scavenging for food alongside the ducks and geese.*

Below: *Tall stands of teasel, thistle and nettle vie for a position near the water's edge, letting loose their seeds to the wind. Many will be carried downstream and dispersed to new habitats, eventually creating new thickets to be colonised by insects, birds and mammals. The green film on the surface of the water is a familiar sight. It is duckweed which grows prolifically wherever the water is enriched with minerals from surrounding farmland.*

melodious chorus of song. Identifying one warbler from another requires plenty of practise since they all tend to look very much alike for the short time that they are generally in view. The three birds most likely to be seen or heard are the reed and sedge warblers and the reed bunting. Bearded tits may also be present, along with the occasional visit from a Ceti's or Savi's warbler.

The sandy banks around gravel pits offer an ideal refuge for sand martins. These skilled fliers are gregarious birds which come back to the same nesting sites each year after their winter sojourn in Africa. They can be seen along with swallows and house martins skimming the expanse of water in their search for mayflies, gnats and craneflies with which to feed their young. Kingfishers also make use of such banks which are ideal sites for their tunnel-like nests.

Muddy areas with plenty of cover can be alive with mallard, teal, sandpipers, water rails and greenshanks, the edges surveyed by the motionless grey heron, and you may even catch sight of a stealthy bittern perfectly camouflaged amongst the reeds.

Literally thousands of birds find a welcome refuge on and around the artificial waterways provided by man, and watching them can

Left: *Toads are much less lively amphibians than frogs because they rely on the poisonous secretions in their skin to deter predators. Common toads and frogs no longer warrant their name 'common' since they are much less widely distributed nowadays, their numbers having decreased rapidly in recent years due to land drainage and destruction of their natural habitats. Garden ponds offer them an ideal refuge and their spawn may be found there in the spring.*

Below: *A true opportunist, the brown rat has learnt to retrieve bread thrown into the water for the fish seen swimming below. An excellent swimmer, this resourceful rodent can swim for 20 metres or more with little problem and colonies of these typically urban and farmyard animals frequently live on canal and river banks.*

provide a fascinating hobby for any bird-watcher. Many such habitats have been designated as nature reserves and good views can often be had without even the need to leave the confines of the car. Moreover, other familiar birds soon come to learn of the places where man gathers and tame blue tits, chaffinches and blackbirds, along with the inevitable sparrows, soon gather in the car parks and along the paths waiting for left over food. As in other places, the magpies, crows and jackdaws are seldom far away, ready to swoop in and claim their share of the scraps!

MAMMALS

Several species of mammal live in and around canals and reservoirs. Some are creatures which only ever live near water, whilst others are more usually associated with the woods or fields but have found a plentiful supply of food in what might seem a quite alien environment.

Brown rats, for example, are quick to exploit any available habitat and the canal towpath is no exception. They are good swimmers and frequently colonize the canal banks, feeding off the scraps thrown to waterbirds as well as their usual diet of seeds and fruit.

Another common rodent is the water vole which frequents watery ditches and canals where it can be seen swimming from bank to bank. Often referred to as water rats, these creatures can be distinguished from brown rats by their shaggier coats and blunter faces. Despite being good swimmers, they can only remain submerged for a short time (around 20 seconds) and can even be found living well away from water, making extensive burrows in soft earth. Their territories are usually limited to around 130 metres of bank and encompass a number of runs burrowed into the bank with entrances both above and below water. Nests of grass are made in which to rear their young, either in the burrow or at the base of waterside plants.

The much-loved otter, which traditionally lived along rivers, streams and by natural lakes, was quick to colonize man-made waterways. But the advent of poisonous chemicals in the 1940s and the progressive loss of their natural habitats through intensive agriculture, has led to their decline in much of lowland Europe. They may still be found along some canals, however. Their droppings (spraints) and areas smoothed down to allow them a 'slide' into the water are both clues to their presence.

Much more likely to be seen is the North American mink. These smaller relatives of the otter were imported to Europe for their fur earlier this century, but many escaped thanks to their ability to climb the often inadequate fencing, and managed to survive in the wild. They are now widespread and their numbers are increasing. Mink are ferocious carnivores that will even enter chicken coops like foxes and kill all the birds. In the wild, they have replaced the otter along the edges of rivers and streams, and around reservoirs and lakes, feeding on fish as well as water birds, voles and shrews. The footprints of both these water-loving members of the weasel family may be found in the mud at the water's edge – sometimes the only indication of their furtive presence.

A very bulky mammal of the waterways is the coypu. A South American native, it was brought to Europe for its fur but succeeded in escaping and setting up colonies in the wild. It breeds successfully in the Camargue region of France, the Broads of Eastern England, and Germany, and would probably be more widespread if it were not for the harsh winters. Weighing over 7kg (20 lbs) with the males often reaching almost a metre in length, these aquatic rodents can cause considerable damage to nearby crops. The paddy fields of the Camargue are frequently disturbed by its extensive burrows which upset the irrigation system, and arable crops such as cabbages may also suffer from their predations. It has proved very difficult to eradicate in those areas where successful breeding has taken place and its footprints, the hind feet only being webbed, can sometimes be found along the banks of slow-moving streams and canals.

MAMMALS	
Bank vole	Field vole
Beaver	Harvest mouse
Brown rat	Mink
Common shrew	Otter
Coypu	Water vole

BANKSIDE MAMMALS

You may only get a fleeting glimpse of mammals along the waterway, so it is best to know what to expect. Many reported otter sightings are, in fact, American mink which are now more common along some stretches of water. Muddy banks may also reveal the footprints of these mammals and droppings can be another useful guide.

Coypu. *These are the largest mammals associated with waterways. They make large holes in banks and are found in the Broads area of Eastern England and in the Camargue, France.*

Water vole. *These small mammals make a distinct plop as they enter the water when disturbed thus giving away their presence. They are good swimmers and can enter their runs from below water level.*

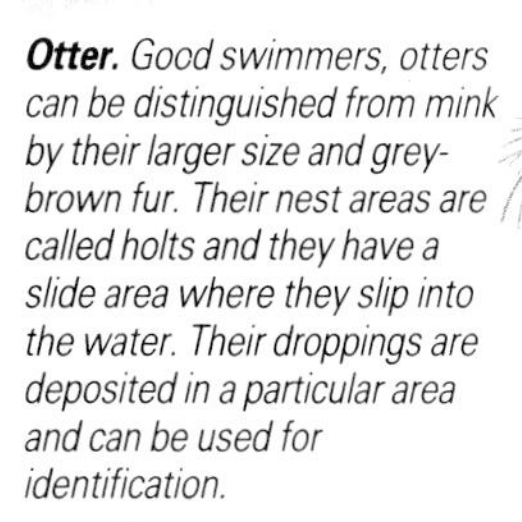

Otter. *Good swimmers, otters can be distinguished from mink by their larger size and grey-brown fur. Their nest areas are called holts and they have a slide area where they slip into the water. Their droppings are deposited in a particular area and can be used for identification.*

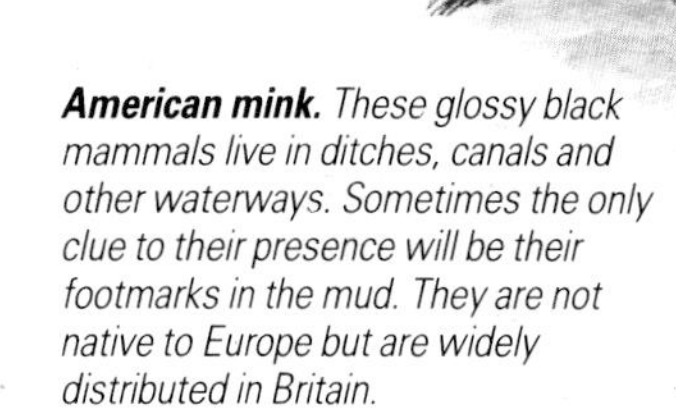

American mink. *These glossy black mammals live in ditches, canals and other waterways. Sometimes the only clue to their presence will be their footmarks in the mud. They are not native to Europe but are widely distributed in Britain.*

Watching wildlife from a fixed hide can often afford the amateur naturalist much closer views than could otherwise be achieved by crouching behind vegetation or sitting amongst the branches of a tree. Such hides are usually strategically placed so that the best views of wildlife can be had over lagoons, moors, marshes and other open stretches of land. All you are then required to do is keep quiet and watch. Double-decker hides are popular and give a higher vantage point. Most hides are so designed that you can approach them without the birds seeing you arrive.

WATCHING WILDLIFE

You don't have to have expensive equipment to watch wildlife successfully. Plants and animals live all around you. To see them, all you have to do is be observant; whether you're going to school, going to work, playing golf or simply taking a stroll in the countryside. Your ears, eyes and nose will be enough, in many cases, to tell you what is about and with a little patience on your part more can be revealed. After all, the last century's great naturalists – Charles Darwin in England, Gregor Mendel in Austria and Jean-Henri Fabre in France – did not have access to modern sophisticated equipment. They just used their powers of observation to help them discover and understand more about the natural world.

NOTES AND NOTEBOOKS

Obviously some items of equipment can be a great boon to the amateur naturalist – cameras and binoculars for example – but perhaps the most useful item is the one most often overlooked: the notebook. Note-taking is a vital part of any naturalist's studies. The living world is so diverse that it would be impossible to remember all the species of plant or animal seen on a single day spent in the country. At the time you may feel you can remember all the birds, flowers and tiny insects but at the end of the day you're more likely to draw a complete blank. Storing the information in a notebook as you go along is one way of overcoming the enormously forgetful nature of the human memory!

There are several different ways of keeping a notebook and you must choose whichever one is most suitable to you. You could opt for a small pocket book in which to record your sightings as you see them, and keep this as your permanent record. Alternatively, this could act as your scribbling pad, your rough notes then being entered into a bigger, more substantial notebook that should be large enough to last for several years.

What sort of things should you put in your notebook? First of all, each sighting should be accompanied by its date and location, the type of habitat in which it was seen – whether a derelict building site, river bank or public garden, and also a note of the weather conditions at the time. You might like to list all the things you see on a walk in the country. Alternatively, you could study the wildlife in one small area – the stretch of road from your house to the nearest post box, for example – and note the seasonal comings and goings. You could also concentrate on a particular group of plants or animals within that area – the number of fungi seen, for instance.

Don't fall into the trap of thinking that you need to be surrounded by green countryside to see something worthwhile – sometimes there's much more wildlife in the town than in the country. Quite a considerable list of birds can be made from such vantage points as high-rise buildings. Similarly, the number of wild flowers sprouting on the verges of a housing estate or local park can be recorded.

If you see something which you cannot identify, it is useful to make a quick sketch of it in your notebook or at least note down any salient characteristics which could help you identify it later on. You might like to take a pocket-size field guide along with you or, once at home, use a larger identification book to help you complete your notes. Many of the larger volumes have 'keys' to help you identify unknown species. They are based on the characteristics of plants and animals and, with the use of illustrations, help you to identify your find by a process of elimination. Some flower books have pages devoted to different coloured flowers – blue, white and so on – to help you in your selection.

Quite apart from looking out for the wildlife, a great deal of information can be gleaned from being a nature detective. Whether you find droppings, food remains or simply footprints, it will give you tangible proof that some creature has frequented the spot recently – and could be likely to do so again. As with all nature-watching, this information will be of greatest use only if you have the patience to sit down and record what you see and think about any significances, similarities and curiosities.

IDENTIFYING FOOTPRINTS AND DROPPINGS

Many mammals can be identified by their footprints – muddy banks near water and snow will often reveal their presence. The criteria to look for are the number of toes, whether a fore or hind foot, the size and whether the animal is walking or running. Remember that not all the toes may make a mark, as in the case of the rabbit (bottom).

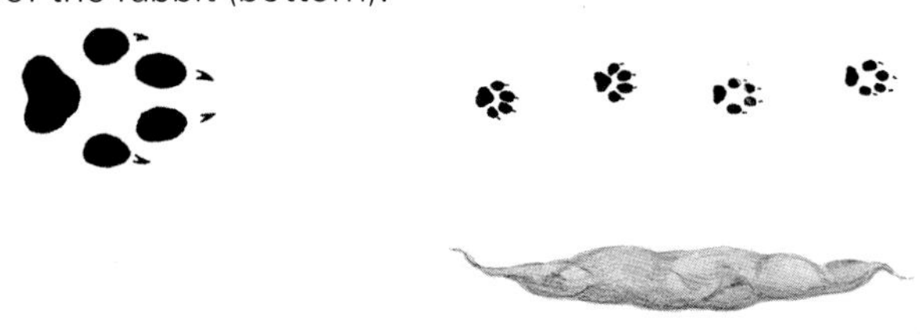

Fox. *Four claw marks and four toe pads. When running (galloping), the neat tracks of the fox appear to be almost in a straight line. When walking (trotting), the footprints appear almost side-by-side. The droppings are greyish and nipped off at one end.*

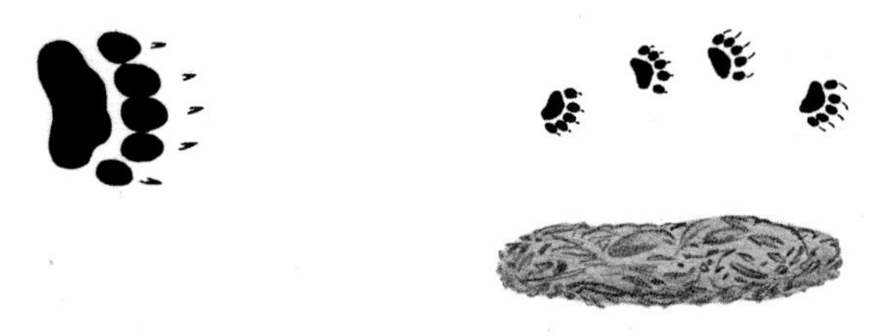

Badger. *A distinct footprint with five toe pads showing long claws. The best places to find them are on banks up which the badgers regularly scramble. Badger droppings are similar in appearance to those of foxes but would be found neatly deposited in a shallow depression or latrine.*

Fallow deer. *Deer all leave similar prints or 'slots', although in some species the dewclaws are also shown. Those of the fallow deer are narrow and pointed. Red deer slots are larger. The black, cylindrical droppings can be used to verify identification.*

Hedgehog. *A five-toed print but, like the fox, only four may sometimes show since the inside toe may be small and too far off the ground to register. The long toes are quite distinctive. The dark cylindrical droppings are often wettish in appearance.*

Rabbit. *Five toes but only four make marks. The hind feet make a long impression which is quite different from the print left by those at the front. When running fast, the prints of the hind feet come before those of the front. The round, fibrous droppings are often left on grassy mounds.*

WATCHING MAMMALS AND BIRDS

To watch animals and birds in their natural surroundings with any degree of success you need to have some understanding of the creatures themselves, a few basic skills and, above all, patience. Apart from a few exceptions, most birds and mammals are wary of man and will not stay around for long once aware of his presence.

As mammals have a strong sense of smell, it is absolutely essential that you position yourself downwind. Basic rules are to move quietly and slowly. If you spot an animal or group of animals, don't walk directly towards them. Instead, approach in a more roundabout fashion as if you are paying attention to something else. Such tricks can often succeed in getting you closer to your quarry.

During the summer, the nests of harvest mice may be found in hedgerows or amongst long grass stems which are split and woven in with growing vegetation.

By understanding a little about the nature of the wildlife under study you will have a much greater chance of seeing it. Most mammals are creatures of habit. They often have regular runs or tracks through the undergrowth, and most have a den or burrow to hide away in during the hours of daylight. Find one of these, employ a little patience and pick the right time of day and you are almost certain to catch a glimpse of some elusive creature.

All birds and animals have to drink so watching a 'waterhole' can also produce results. You may see swifts or martins winging low over the water, skimming its surface for refreshment. Magpies, pigeons and other smaller woodland birds may come to the water's edge to drink, and the surrounding mud may well reveal the footprints of visiting nocturnal mammals – mink, foxes, badgers, even the odd weasel.

Since most European mammals are nocturnal, the hours of darkness are obviously the best time to go out looking for them. Serious naturalists are often aided in this potentially toe-stubbing task by the use of infra-red attachments to cameras Most nocturnal mammals cannot see this wavelength and remain unaware that you are watching – unless their other keen senses give you away.

Wherever and whatever you are watching, whether it's the nest of a warbler or the antics of young fox cubs, be careful not to disturb your subject. Birds will desert their nest and eggs and mammals leave their burrows if they think they are being threatened by some predator. In all cases, it should be the well being of the wildlife that is of paramount importance.

Rather than attempting to watch creatures in their natural habitat, another alternative is to entice them to come to you! Bird trays, nut bags, and nest boxes for tits, owls and bats quickly

Weasels often investigate the runs of other animals, being small enough to follow mice and moles along their tunnels. If you see one disappearing into a burrow, wait quietly and it may well reappear.

increase the wildlife potential of your garden, however small. Similarly, a wealth of flowers growing in the herbaceous border, or a wild corner rich in nettles and thistles will always provide a haven for insects. Moreover, whatever you do to conserve wildlife in your garden will help to increase the wildlife potential of your local area. A more detailed account of what can be done is covered in the following chapter.

RECOGNISING SIGNS OF MAMMALS

Many mammals are active only at night and spend much of the day asleep. Therefore, we tend not to see them very often. We may be rewarded by the glimpse of a field mouse or weasel frantically running across the road when we are driving along at night, or the sight of rabbits grazing by the side of the motorway, oblivious to the hum of cars passing so close by. On the whole, however, our native mammals – with a few exceptions – use their excellent senses to help them keep well away from our prying eyes.

The tell-tale signs of their existence are somewhat easier to find however, and can often lead to sightings of the creatures themselves. If you know for certain, say, that a hedgehog is regularly using your garden as a nocturnal hunting ground, then patiently waiting for it to

BIRDS AND THEIR BILLS

The shape and size of a bird's bill can tell you a lot about the food it eats. There are long, thin bills – like those of many waders – for probing and picking up food from difficult places, all-purpose bills for omnivorous appetites like those of gulls or rooks, and fat chunky bills for cracking nuts.

Apart from size and shape, colour can also be important. The bright blue, yellow and red stripes of the male puffin's bill, for example, are used to attract females during the breeding season. The bills then become duller during winter.

Goldfinch. *A typical seed-eater, this bird has a pointed bill for opening up the buds of dandelions and the fruiting heads of burdock, knapweed, thistle or teasel. It is then used to pick out the seeds.*

Bee-eater. *The long, slim bill of this bird is slightly curved downwards. It is used to catch dragonflies, damselflies and hoverlies, the stings of which are removed by the bird before eating. It also uses its beak to tunnel into sandy banks where it makes its nest.*

Mallard. *A typical duck's bill – wide and flat. It is very useful for filtering out small plants and animals from the water which is taken in and forced out of the sides of the bill through tiny grooves which trap food particles.*

Redshank. *A long, fine-pointed bill which is ideal for probing mud and shallow water. Used in a forceps-like manner to catch fast-moving invertebrates such as lugworms, as well as small shellfish which are then prised open.*

Kestrel. *A hooked bill typical of predatory birds like hawks and owls which kill small mammals. The bill is used to tear flesh and fur from the victim's body.*

Hawfinch. *A bulky bill used like a pair of nutcrackers to crack open fruit stones, seeds and nuts. With its powerful muscles it can exert a force of 60-95lbs to crack the most stubborn of nuts. Cherry stones are its favourite food.*

ANALYSING FOOD REMAINS

With a little careful observation, you can find the food remains of many mammals, birds and even insects. Stripped fir cones, broken shells, plucked plumage, pellets and birds' droppings revealing seeds are things to look for. Bird pellets can contain all sorts of plant and animal remains – indigestible food regurgitated from the gizzard. Owls, crows, gulls, jackdaws and oystercatchers all bring up pellets.

Along a path in Southern Europe you may discover a collection of butterflies' wings discarded randomly beneath a thistle plant. Closer inspection may reveal a praying mantis lurking amongst the flowers, waiting for another victim to come visiting the plant for nectar.

Pine cones are stripped differently by small mammals. They remove the hard protective scales in order to get at the nutritious seeds. Mice generally make a neater job of it than squirrels who leave a ragged cone.

The caps and stems of many species of fungi are eaten avidly by slugs. Their tell-tale signs are clearly obvious since they eat deep into the cap, leaving the paler insides showing. Slugs will also venture onto animal droppings to eat the discarded plant fibre.

The pellets regurgitated by carrion crows differ from owl pellets in that they do not contain as many bones. They are about 2cms across and up to 4.5cms long, containing matted dark fur and chaff in the autumn.

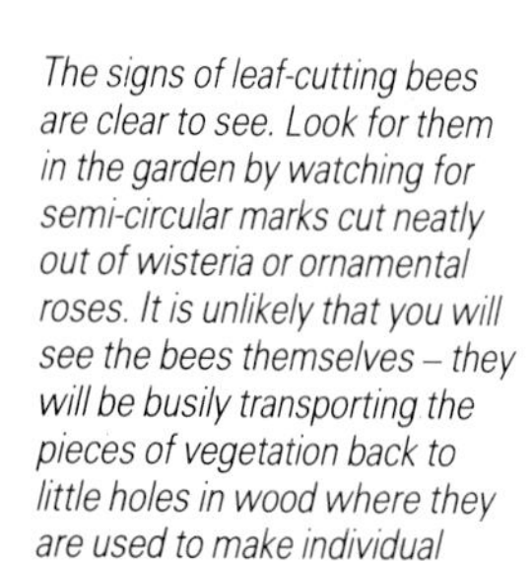

The signs of leaf-cutting bees are clear to see. Look for them in the garden by watching for semi-circular marks cut neatly out of wisteria or ornamental roses. It is unlikely that you will see the bees themselves – they will be busily transporting the pieces of vegetation back to little holes in wood where they are used to make individual cells for their larvae.

appear will probably yield results.

So how do you know what mammals are around? Among the first things to look for are droppings. Many of these are not easy to tell apart but, given other relevant information – the type of habitat, the time of year, the rarity of the creature and so on – certain assumptions can be made. In the house, the small, dark droppings of the house mouse are enough to give away its presence. Gnawed cardboard boxes in the larder or shed, spilt food on the kitchen floor and chisel-like marks on soap bars or plastic are also tell-tale signs of mice. With rats, the damage caused by their night-time excursions is likely to be much greater – they have even been known to gnaw through bricks to reach food supplies. If you have a colony of bats in your attic or cavity wall you may find a few of their long black droppings scattered on the ground outside. Unlike other small mammal droppings, those of bats are dry and fibrous. If they are inspected carefully they will be seen to be composed of the crushed remains of insects, usually beetles' forewings which are indigestible.

Fox droppings are often found in gardens and on wasteland. They are usually greyish in colour and are nipped off at one end. With age they fade

Tell-tale signs of a badger's presence can be seen in the form of scratch marks on trees growing near the sett. Badgers use the rough bark of all sorts of trees to sharpen their claws.

to white. Foxes often use some of the footpaths and tracks made by man, and the droppings are frequently found there. Badger droppings are similar in appearance but would not be found in the same situation. They are deposited in a proper latrine area near the sett. A shallow depression is scraped out of the ground and the wet droppings left there. Unlike domestic dogs or cats, no attempt is made by the badger to cover them up as their smell is used to scent mark the creature's territory.

Rabbit droppings are a fairly common sight in grassy areas. Their communal toilets are often found on raised patches – grassy tussocks, the tops of ant hills and so on. In the garden the cylindrical droppings of hedgehogs may be seen on the lawn or patio. Clues to the presence of mammals can also come from their fur. The characteristic hairs of badgers and foxes can be seen caught up on barbed wire, and clumps of rabbits' fur are often left on the ground by these creatures after grooming.

Finding the entrance to an animal's living quarters can also provide you with an ideal spot to watch if you want to catch a glimpse of the owner. The first problem, however, is to decide whether the burrow is still in use. Dead leaves,

Breaking and entering nuts is a specialized art. The top left hazel nut has been eaten by a bank vole leaving a neat round hole, the nut in the centre has been neatly split in half by an adult squirrel and, like the bottom right example, shows the exit hole of a nut weevil.

twigs and other vegetation cluttering the entrance will usually indicate that it has been left empty for some time, but if it is clear and the vegetation in front flattened down or pushed aside then the owner may well emerge at dusk. A second problem is to decide to what animal the burrow belongs. This can often be difficult but size, location and habitat are the main factors.

Apart from footprints (see page 144), food remains are another good indication of what small mammals are living in an area. Discarded hazel nut shells, for example, can often be used to give a clear indication of the mammals and birds present without them actually being seen. Squirrels, wood mice, bank voles and woodpeckers all tackle nuts in a different way to get at the protein-rich kernel and leave the evidence of their presence on the shells. The same applies to fir cones.

WATCHING INVERTEBRATES

With many invertebrates, finding the creatures themselves presents little problem. Lift any cascading clump of vegetation and a myriad of small creatures will scuttle before your eyes. Weed or dig any patch of an herbaceous border and you will doubtless be confronted by beneficial earthworms, millipedes, shield bugs in their clumsy armour and countless aphids, ants and other insects.

BADGER-WATCHING

Watching badgers is perhaps one of the most pleasant pursuits to be enjoyed both in the town and country. These striking creatures live in well organised family groups of up to 15 individuals, in setts often dug on escarpments where bands of rock are exposed. They thrive in wooded suburban areas, in woods and copses and around parks, commons and on railway embankments. Thanks to the increasing urban sprawl many badgers find themselves encircled by housing developments and many home owners are fortunate enough to have badgers visiting their gardens, searching for scraps. They love sugary breakfast cereals.

You can find out if you have badgers living nearby by looking for their well-trodden tracks up and down slippery hedgerow banks or their black and white hairs caught on barbed wire fences. You might find their sett which is an ideal spot to watch if you want to catch a glimpse of the creatures themselves.

Badger-watching is best done in late April and early May when the cubs will be allowed above ground. It is best to approach the sett as quietly as possible about one hour before dusk to choose your position. This should be downwind of the sett and perhaps up a tree since badgers, like humans, are less likely to look overhead for signs of danger. As twilight approaches you may see a black and white head

Badgers excavate tonnes of soil and stone – often on ancient sites that have been continuously occupied by badgers for hundreds of years – to form lengthy passages radiating from a communal chamber. These setts can be easily distinguished from foxes' earths since they have clean tidy entrances and a system of well-trodden paths.

gingerly protruding from one of the holes, the nose sampling the night air for any foreign smells. You must not stir or make a noise at this moment since the badgers will be alerted and will not come up until much later. If all goes well, you may be rewarded by the sight of a family of badgers grooming themselves or simply playing before heading off on their nocturnal hunt for food. On other nights you may see only a single badger and eight out of ten times you may not see anything at all! As we have already said, the greatest asset of any naturalist is endless patience.

PITFALL TRAPS

One way of studying nocturnal insects is to make a pitfall trap in which to catch them. You can make your own version very simply by sinking a yoghurt pot into the ground so that its rim is flush with the ground surface. To make a pitfall trap, pack the earth well up to the edge and ensure that there is a hole in the bottom so any rain water can drain away. The best place to dig these traps is in a wood or under a hedgerow, though one positioned at the edge of the garden or in a flower bed can reveal some unexpected finds.

Having set the trap at dusk, you may like to return a few hours after darkness or wait until morning to find out what has been caught. Many nocturnal insects and invertebrates will have fallen into the pot and been unable to escape.

To study your captives, a hand lens will come in very useful. They are sold in most good camera shops and can be mounted either in plastic or metal. They vary in magnification from x4 to x15 (x10 is a useful size to have). Hand lenses are useful for inspecting the delicate structures of flowers as well as many small creatures. There will be plenty of 'minibeasts' that might fall into your trap that are difficult to see clearly with the naked eye and therefore need to be observed with the aid of a hand lens. If the lens is tied around the neck on a lanyard it is easily accessible and will not be so easily lost. Magnifying glasses are more cumbersome to carry about, although there are some which fold away nicely into the pocket. Large solid magnifying glasses are good to have at home however, so that seeds, shells, insects and other 'finds' can be inspected at leisure.

Such equipment will enable you to count the legs of smaller invertebrates – a necessary part of their identification. Beetles (with six legs), spiders (eight legs) and woodlice (fourteen legs) are likely finds, so are centipedes (with over a hundred legs) and millipedes (with several hundred).

The caterpillars of moths and butterflies can sometimes be detected by their droppings or partially eaten leaves. Because of this, some moth caterpillars go to elaborate lengths to flick their droppings away from their feeding area so that predators are not given an easy clue to their whereabouts. If you walk through an oak wood in May or June you might mistake the sound of droppings falling from the canopy for the gentle patter of rain. This is due to tens of thousands of different caterpillars which feed on the oak leaves. Occasionally they will succeed in stripping the leaves bare, and if the ground beneath is carefully inspected it will be found to have a fine covering of droppings. Similarly, any car parked under such trees will soon have a roof peppered with black.

Later in the summer any car left under sycamore or lime trees may be covered in a sticky

Leaf mines, like this on bramble, occur on many plants. They are the traces left behind by the caterpillars of micro-moths or tiny flies which feed between the upper and lower leaf surfaces.

substance that is difficult to remove. This is honey dew – the liquid droppings of aphids. Aphids feed on the plant sap tapped from the leaves and stems and pass out the excess sugary solution which drips down from the trees. Because of its sugar content, honey dew is frequently gathered up by butterflies, honey-bees and foraging ants.

EQUIPMENT

Although you don't *need* any equipment to watch wildlife, there are a few items that can greatly increase your enjoyment of the subject as well as making life considerably easier.

Binoculars are perhaps the most essential piece of equipment for many naturalists. They allow you to observe all kinds of creatures without unduly disturing them. They enable you to bring birds and animals close enough to distinguish their shape and colour, and also give the user an opportunity to watch such creatures going about their daily business, unaware that they are being watched.

But which ones do you choose? In the shop,

the detail to look for is the magnification. Each pair of binoculars will have this shown on the side as, say, 7x50. The first number is an indication of the degree of magnification, ie a bird seen at a distance of 700 feet would appear

These leaf-blade galls harbour the larvae of a tiny gall-midge. When mature, the larvae leave the gall and pupate amongst the leaf litter. The galls turn different shades of yellow, red and purple as autumn advances.

to be 100 feet away. Here, a figure of between 6 and 10 is the one to aim for – less than 6 will not bring the subject close enough; more than 10 means you'll only see a small part of the whole scene through the binoculars and may need a tripod in order to use them effectively.

The second figure indicates the width of the objective lens in millimetres. The bigger the objective, the greater its power to collect available light at twilight and dusk. This is a useful facility for naturalists, particularly those who want to watch badgers or foxes that will only appear at such times. Aiming for a value of about 40 to 50 will allow you to see better when your own eyes fail to pick out objects after dark, but the price you have to pay for this is a bulky and heavy pair of binoculars.

In towns and cities you usually want to see across short distances rather than across great stretches of moorland. You may have a bird close by that you want to study in greater detail so in this case you need a pair of binoculars which focus down to a few metres. There are some very slim-line binoculars of moderate price available which would be very useful for this.

Photographing wildlife can be an absorbing hobby in itself. It is a great method of keeping a record of your findings and the thought of a resulting photo often acts as an added incentive to anyone patiently waiting for or stalking a subject. For this reason a good camera is a welcome piece of equipment for many naturalists.

The popular fixed lens cameras have limitations for the amateur naturalist. They are ideal for taking pictures of habitats such as derelict sites, cemeteries or reservoirs but they are not suitable for close-up pictures of wild flowers, insects, birds or mammals. The serious amateur needs to buy a single lens reflex (SLR) camera which can take a variety of interchangeable lenses. This enables both habitats and close-up shots to be taken. A wide angle lens is very useful for habitat shots in the urban environment where room is severely restricted. It can also be used to take group shots – deer feeding in parkland, for instance, or the waterfowl on a local pond.

For close-up shots of birds, mammals and insects a telephoto or zoom lens is ideal. A 80-200mm zoom is the most versatile as it can be used to take a close-up and then opened out to include more of the background scenery, thus eliminating the need for you to walk back and forth to set up a picture. A set of three extension tubes used in conjunction with a 50 or 135mm lens is useful for close-ups of insects and flowers, but for real close-ups you'll need a macro lens of about 100mm. With any lens larger than 200mm you'll need to use a tripod to eliminate hand-shake. The same applies to shots taken at dusk using a long exposure.

Many people prefer to take photos using natural light but in many cases some sort of extra illumination will be necessary. Automatic flash

Small portable hides can be put to effective use in the garden as an aid to photographing and studying birds. They will become used to its presence and venture within 1 or 2 metres of the avid watcher.

guns are readily available and, if used sensibly, will not unduly disturb the subject. It is often useful to have two small flash guns – mounted either side of the camera to give equal light. One can be rigged up with a 'slave unit' – an electronic device which automatically sets off the second flash when the first one is triggered by the camera.

Serious naturalists may find the use of a hide invaluable in enabling them to study wildlife at close quarters and they can be home-made. They can provide the *only* way of getting close to subjects like nervous birds.

Conservators can undertake valuable work, clearing and improving natural habitats. Here students help clear a small stream, hoping to re-establish a rich aquatic habitat which will become colonized naturally by all manner of plants, insects and wildfowl. Ponds always fill up with debris and the presence of crack willow helps further in drying up the habitat.

CONSERVATION

People often ask "What is conservation?" The answer is that it has now become a very popular branch of science which seeks to protect plants, animals and their habitats. It represents a new awakening in people – a concern about the state of the countryside in which they live and the welfare of its wildlife; an expression of their urge to respect their wildlife heritage.

Conservation did not always stand for wildlife, however. It was formerly more associated with historical conservation – the upkeep and preservation of tythe barns, listed buildings and the like. Now that there is a greater awareness amongst the public, thanks mostly to television and radio, conservation has begun to assume a new mantle – that of the protection of wildlife and its habitats, from the rainforests of New Guinea, to ancient woodlands in Europe or the peat bogs of Southern Ireland.

But where would the conservation effort be without people? So much conservation depends on thousands of people working voluntarily in a part- or full-time capacity. A continual war is waged against authorities to protect threatened habitats, spare interesting corners, stop road-building and counter pollution – all in the name of wildlife. It has to be said, however, that the need for conservationists would not be necessary if it weren't for people – we only have ourselves to blame. We demand the electricity, the water, the transport systems and the expansion of our towns and factories. We must accept then that, hand in hand with this, will come a country filled up with reservoirs, pylons, power stations, houses, factories, dumps and motorways. Our greatest problem is that there are too many people living in a constricted environment. The more people, the more endangered becomes the state of the countryside. So in overpopulated countries natural habitats and wildlife are many times worse off than in countries where there are more open spaces to survive in. But that still doesn't mean that wildlife is safe. How we conserve it is a matter of choice. In Western Europe many countries use a criterion which appears, on examination, to be absurd. However, it can be explained easily.

Where there is a large population enlightened into the ways of conservation, concentrated into a small space (Britain, for example), the need to conserve wildlife and habitats becomes very urgent. The more man expands into the countryside, the more wildlife contracts. Therefore, the emphasis on conservation is that we should try to conserve those rare plants, animals and habitats that we discover. If we find a meadow full of orchids or butterflies which we want to conserve as it is, it means that we have to try and defy the natural changes of nature. *Habitats always change* – meadowland is succeeded by scrub and then woodland in a natural progression – so this philosophy makes mountains of work for conservationists from the outset. Working parties are needed to bash scrub, voluntary muscle to clear invasive grasses, shrubs and trees, yet many countries base much of their conservation on this philosophy. Others take a simpler path...

Where there is more space available for wildlife – in the larger countries of Southern and Central Europe for instance – the criteria may be different. A clearing full of orchids and butterflies can be left to grow up naturally into woodland without any trace being left of the interesting flowers and insects that once existed there. However, there is no loss to the conservationists because similar clearings will soon develop elsewhere in the forest – through the death of large trees, forest fires etc. In other words, if there is sufficient space, nature will conserve itself.

HOW TO CONSERVE

To conserve any habitat successfully you must first know what is present there; to conserve any plant or animal you must first find out how it lives. In other words, a lot of initial ground work is required before conservation on any level can begin. If you are hoping to conserve your local pond, woodland or wilderness you must first study it very carefully. Only then can you begin to conserve your favourite habitat and wildlife.

On a wider scale, you may like to take part in a

UNDERSTANDING PLANT SUCCESSION

In studying any habitat, you must realise that it will eventually change – not through the hand of man, although this may also happen, but through natural plant succession. In five years' time, pleasant grassland may have developed into light woodland or scrub and, if left to continue unhindered, a forest would eventually result.

As you travel around the countryside look out for the different stages of encroachment of scrub on meadows and pastures.

This grassland habitat is made up of about 90% grass and only 10% scrub. It will support typical grassland species such as rabbits, field voles, orchids, many species of butterfly and other insects. There will be few shrub species such as hawthorn, sloe or wild rose, known collectively as scrub.

If grazing animals (whether domestic or wild) are removed from the area, small bushes will develop, including examples of hawthorn, wayfaring tree and sloe, along with thickets of bramble. After a couple of years, the habitat will have progressed to one containing 90% scrub and only 10% grass.

After five years, the same area will have developed into light woodland. Scrub trees such as silver birch, elder, oak, ash, holly and hawthorn may be seen, along with the remaining impenetrable bramble thickets. A new set of wildlife will also be in residence – woodland birds will sing from the treetops and larger mammals make their dens in the undergrowth.

national survey of wildlife and habitats. Several of the voluntary conservation and wildlife bodies carry out such surveys. Ones such as tree surveys of towns and parks, wildlife of cemeteries and invertebrate site surveys make interesting projects to become involved in. You will discover a lot which may lead you into other areas of fascinating research. Simply exploring your local countryside may also reveal forgotten corners worthy of study and conservation.

Conservation groups often run field study days, encouraging members to become involved not only in helping to conserve habitats but in studying and understanding more about the natural world. Here students take a break from their survey of wild plants using quadrats.

CONSERVATION AT HOME

Although the initiative for many projects often originates with national conservation bodies, there is a lot that can be done by the individual to help wildlife – in the garden. By creating inviting habitats for wild plants and animals on your doorstep, you are increasing the reservoir effect of gardens as sanctuaries for wildlife. This does not mean that you will have to do a lot of hard work – quite the opposite in fact. A corner left to grow wild with grasses, nettles, thistles and other 'weeds', a wall left to grow thick with ivy and other creepers, and perhaps a bird box or two can make all the difference.

BIRDS

Lots of people already encourage birds. Many gardens sport bird tables and boxes, and some people even go so far as to give out several hundredweight of bird food each year. This behaviour is undoubtedly of great benefit to the visiting birds, particularly during hard winters, since many would die of starvation and cold if help of this kind were not at hand.

The provision of a bird box or table also gives you a wonderful opportunity to study birds more closely. A table, sensibly situated and made so that the birds do not become easy prey for cats and other predators, can be used to attract many

AUTUMN FRUITS FOR BIRDS

Apples	Oak
Blackberry	Raspberry
Cherry	Rosehips
Cotoneaster	Rowan
Guelder rose	Snowberry
Hawthorn	Sunflower
Holly	Teasel
Honeysuckle	Wayfaring tree
Mulberry	Yew

species – from rare migrants to common natives. Household scraps and seeds can be used to attract finches, blackbirds, robins and sparrows. Similarly, the bird life in your garden may increase with the provision of nest boxes. A great variety can be bought from garden centres, pet shops or by mail order, and simple boxes can be made at home from scrap wood. The two most crucial factors in determining the success of a nest box are the size of the entrance hole and the position of the box itself. The panel opposite gives further details.

You can also garden for birds. Providing

When clearing light woodland, many conservators leave tree stumps as posts which act as a refuge for many kinds of wildlife. Woodpeckers will be attracted by the sound of insects living within the posts and peck out holes in order to reach them. Willow tits then excavate the holes further and use them as nesting holes.

plenty of fruity shrubs such as cotoneaster, hawthorn and privet as winter food for both migrants and residents is one aspect. The other is to leave hedgerows and shrubs to thicken out, thereby providing further natural nesting sites. Ivy-clad walls are also useful in this respect. A supply of water will also receive grateful visitors, as will windfalls left – not only for butterflies and honey-bees – but for birds to feed on.

MAMMALS

There are probably quite a few resident mammals in the average garden, although you may not see them. Large creatures such as foxes and badgers have a tendency to roam through suburban gardens on their nocturnal wanderings, so if you provide food for other

MAKING NEST BOXES

Bird and mammal boxes can be made from offcuts of wood. Their positioning is important and also the size of the entrance hole. Don't be surprised if dormice, wasps or bumble bees take over your tit box – they will find the accomodation to their liking as well!

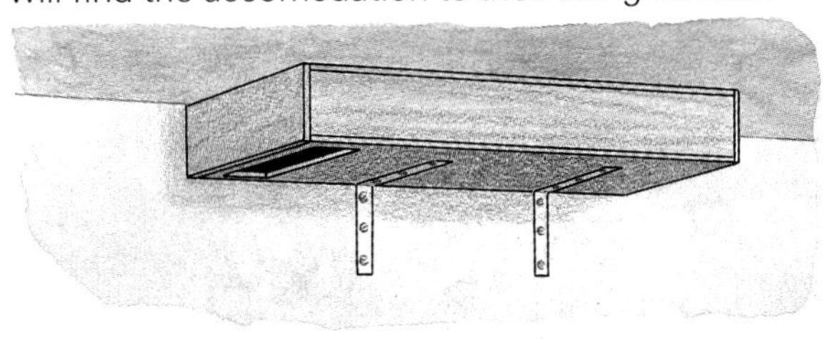

Swift. *These birds like a long box with an entrance on the bottom surface measuring 150 x 40cms. It should be fitted high up under a bridge, the eaves of a barn or similar structure to allow the free flight of this swooping bird. Dimensions: 15 x 10 x 40cms. Hole: 10 x 5cms.*

Kestrel. *A sturdy box is needed here, with an open front for perching. It should be fitted high up in a tree, out of direct sunlight. Dimensions: 30 x 30 x 60cms.*

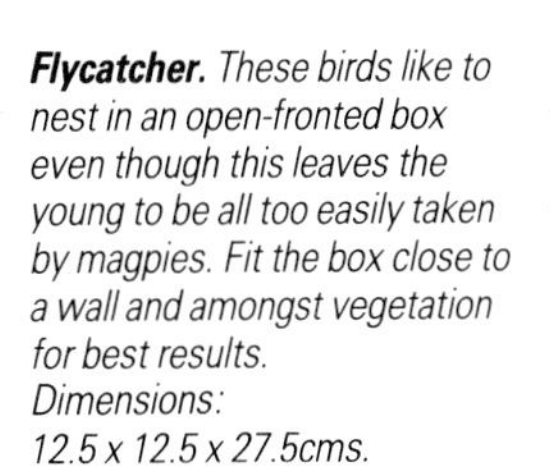

Flycatcher. *These birds like to nest in an open-fronted box even though this leaves the young to be all too easily taken by magpies. Fit the box close to a wall and amongst vegetation for best results. Dimensions: 12.5 x 12.5 x 27.5cms.*

Blue tit. *Tit boxes have special openings of the correct diameter (30mm) to allow only the blue tit to enter, although other small birds such as great tits, tree sparrows or nuthatches may also gain access. Place them on the north side of a tree (out of the glare of the sun). Dimensions: 12.5 x 12.5 x 27.5cms.*

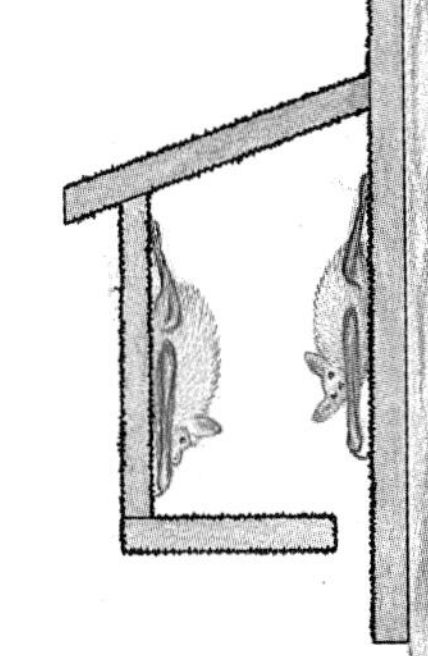

Bat. *Bat boxes are made out of roughened wood with a slit entrance on the underside which should be no more than 2cm wide. The grooves in the wood (which can be made with a saw) are to allow the bats to get the best purchase for walking around. The box should be positioned high up on the south side of a tree to get the warming benefits of the sun in summer. Dimensions: 10 x 10 x 20cms. Entrance: 2 x 10cms.*

CONSERVATION AT WORK

A widespread concern for the well-being of wildlife and its natural habitats has stirred many people into active work – restoring amenities, providing urban sanctuaries for birds and mammals, digging and dredging ponds and canals, forming nature reserves out of overgrown cemeteries or deserted wasteland, or simply by allowing a part or the whole of their garden to go 'wild'.

Such conservationists can rely on nature's amazing ability to colonize new ground, often against all odds, so that a wasteland site will spring up with a host of opportunist wild plants, hordes of interesting invertebrates and numerous species of bird almost overnight. Conservation groups and community spirit are at last helping to give wildlife a chance in an otherwise hostile environment of urbanization, industrialization and sterile agricultural land.

1 *Greater reed mace*
2 *Rooks*
3 *Robin*
4 *Cramp balls fungus*
5 *Winter heliotrope*
6 *Wood anemone*
7 *Water lily*
8 *Coltsfoot*
9 *Mallard (drake)*
10 *Mallard (hen)*

wildlife in your garden you can expect these creatures to sniff it out too.

Small mammals such as mice, shrews and voles are best encouraged by leaving long grass. These animals like to live out of the gaze of predators below the tangle of matted stems. Similarly, sheets of corrugated iron or wooden boarding left lying on the ground may soon shield the nest of a field vole or wood mouse as well as a multitude of invertebrates.

Hedgehogs can be encouraged into the garden by providing a lean-too shelter for their summer nests and hibernation sites. A board leaned against a wall or shed, the space between filled with straw, makes a useful resting place for these beneficial mammals of the garden. Bats can be encouraged by the making of bat boxes, fixed high up on the south side of buildings or trees.

The provision of food for mammals is best done by encouraging the invertebrates and fruits on which they feed. Berries such as hawthorns and rose hips will provide a feast for small mammals such as shrews and mice, and milk may be left out for hedgehogs, but a garden rich in insects will provide the greatest attraction to many.

AMPHIBIANS AND REPTILES

Both amphibians and reptiles have had a particularly hard time in the agricultural countryside over recent years. Thanks to man, the garden habitat offers a refuge for their survival. The greatest contribution you could make to this sector of European wildlife would be to build a garden pond. It not only acts as a focus for amphibians but provides a place where aquatic plants can thrive as well as insects such as water beetles, dragonflies and damselflies.

If you provide the water, much of the resulting aquatic life will come by itself. You can assist in its colonization by introducing spawn from ponds with excess, and perhaps a selection of aquatic plants. The rest will happen naturally.

Even without a pond, frogs and toads may find refuge in the damp, shady areas of your garden – in wet, long grass, in the rockery or hedgerow ditch. Long grass encourages the slow worm – a legless lizard – as well as the grass snake which will feed on the small mammals it finds living there. Walls and patios offer plenty of warm spots for basking lizards.

INSECTS

There will be many uninvited insects already living in your garden. They will not all be cashing in on the fruits of your labour however – many will be thriving in the hedgerow, the long grass, the compost heap and other odd corners. Of all the creatures in the garden, insects are the easiest to encourage and conserve. Just as with birds, you can garden for insects, providing plants and habitats which they both need and like.

Countryside butterflies can be attracted into the garden by the planting of suitable nectar sources. *Here a silver-spotted skipper, feeds on the prolific amount of nectar from a colourfil zinnia.*

Butterflies are colourful visitors and can be encouraged to stay longer if nectar-rich plants are present. Buddleia, the butterfly bush, is undoubtedly the most attractive but Michaelmas daisies and ice-plants will often attract hordes of small tortoiseshells, peacocks and red admirals in the autumn. If you want butterflies to breed, why not just leave some of your garden to go wild. It may well spring up with just the right caterpillar food plants without any of your help.

Piles of old logs and tree trunks offer a refuge for countless different types of beetle. Many are wood-boring beetles whose larvae feed inside the decaying timber for two or three years, so helping in the process of decay. Fungal spores then travel down the beetle tunnels and work away at the wood from inside. Birds, such as wrens and titmice, will be attracted to the wood in search of the insects and their larvae, so a single tree trunk can easily support a whole community of wildlife.

NECTAR RICH PLANTS FOR BUTTERFLIES	
Aubretia	Lavender
Coltsfoot	Lemon verbena
Bramble	Lesser celandine
Buddleia	Michaelmas daisies
Fleabane	Mignonette
Globe thistle	Privet
Honesty	Radish
Ice-plant	Red valerian

WILD PLANTS

Conserving interesting wild plants in your garden can increase their overall population in your locality. You may wish to create a corner of old meadowland by planting poppies, cornflowers or ox-eye daisies, or simply introduce a rare tree or shrub. Before you start however, it is always best to find out what type of soil you have – whether acid, alkaline or neutral – then you can decide which plants will best grow there.

Many people like to take advantage of the packeted seed now sold in shops. You can buy seeds of traditional wild flowers which you may never now see growing on the waysides and in the woodlands of your own country. By growing them in the garden, you can help to increase the incidence of these rare plants, many of which are quite attractive enough to find a place among their cultivated relatives in the herbaceous border.

Opening up your garden in this way, making it a place to be colonized by all sorts of creatures both big and small, can only be for the good. Instead of a relatively sterile environment of neat lawns and rigorously weeded flower beds, you can cultivate a garden that is not only pleasing to the eye but is also teeming with all manner of wildlife. There will be nesting birds to study in the spring, a host of butterflies to brighten an already colourful border and the occasional chance to watch a hedgehog on its night-time wanderings. By looking at your garden as a sanctuary for wild creatures you can play your part in helping to reinstate the lost wildlife heritage of the countryside, and make your own contribution to the conservation of the living world on which man has had such a devastating effect.

NECTAR RICH PLANTS FOR MOTHS	
Cow parsley	Privet
Honeysuckle	Ragwort
Knapweeds	Scabious
Lemon verbena	Soapwort
Night-scented stock	Tobacco plant
Pelargoniums	Valerian

FOOD PLANTS FOR COMMON CATERPILLARS	
Bird's foot trefoil	Common blue
Buckthorn	Brimstone
Cabbage	Large white, small white
Gooseberry	Comma
Grasses	Gatekeeper, small heath, speckled wood, ringlet, small skipper
Lady's smock	Orange tip
Nettle	Peacock, red admiral, small tortoiseshell

INDEX

All entries in italic refer to illustrations.

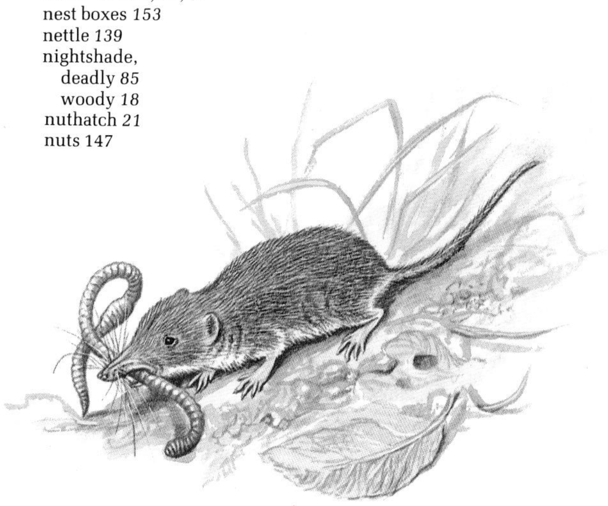